BASIC GUIDE
TO
FLYING

BASIC GUIDE
TO
FLYING

Paul Fillingham

Hawthorn Books, Inc.

Publishers/New York

Line drawings by Alan Maresca

PHOTO CREDITS—Aero Products Research: pp. 14 (top), 19; American Airlines: pp. 125, 127; Bede Aircraft: pp. 35, 166; Beechcraft: p. 155; Boeing Aircraft: p. 132; British Information Services: p. 261 (top); Cessna Aircraft: pp. 14 (bottom), 63 (top), 120, 140, 187, 278, 279; Mr. Willard Custer: p. 262; De Havilland Aircraft of Canada: p. 260 (bottom); Eastern Airlines: p. 267; Edo-Aire: p. 21; Paul Fillingham: pp. 8, 13, 28 (bottom), 40, 42 (both), 112, 150, 184, 261 (bottom), 263; Garrett Airesearch: p. 107 (both); Genave: pp. 223 (middle), 225 (bottom), 226, 250; Grumman American: pp. 2, 5, 28 (top), 146; Narco: pp. 223 (top & bottom), 225 (top), 230; Oldtown Canoe, Corp.: p. 121; Piper Aircraft: pp. 58 (top), 119, 143, 153; *Plane and Pilot Magazine*: p. 109; Rockwell International: pp. 128, 134; Safe Flight, Inc.: p. 18; Mr. Art Scholl: p. 285; Schweizer Aircraft: p. 115; Smithsonian Institution: p. 276 (top); Sturgeon Air Limited: p. 173; Mr. John W. Taylor: pp. 260 (top), 276 (bottom); Mr. Molt Taylor: p. 169; Tetarboro Flight Service Station: p. 85; Type Systems, Inc.: p. 92.

BASIC GUIDE TO FLYING

Library of Congress Catalog Card Number: 73–365

ISBN: 0–8015–0526–7

4 5 6 7 8 9 10

Contents

Preface

Books of instruction without due acknowledgment seem to have come in—well, as far as Western "civilization" is concerned—with the Gospels and, one hopes, went out with the Pentagon papers. This preface is a way of saying thank you to the following:

Bill Byshyn, who labored long as technical editor.

Otis Hardy Maclay III, instructor-general and Grand Inquisitor.

First Officer Glen Nevil-Smith, BOAC–British Airways, who very kindly read the original and made numerous excellent improvements.

And Bob Parke, Bob Stangerone, and John W.R. Taylor.

Flying is something most of us only get to dream about. In most countries of the world it is highly regulated, quite possibly because it is so freeing. And bureaucracies tend to be terrified of public freedom.

In our world the restriction is upon an economic basis since, for officialdom, this is the most acceptable way of measuring competence. However, we don't need to measure this way about flight. We need to build skill. We need to process out those who don't have skill, not those who don't have money. This book, therefore, is really for those who think they're about ready to discover flight. Those who are not frightened to extend their horizons beyond their societally and familiarly self-limiting programs—which includes most of us, I'd guess, these days.

Since there are no new heresies to be unleashed herein, our supporting cast deserves its share of thanks. Alphabetically these are: M. Aasland; I. Allen; Barbara Burg; M.B.; R.B.; Ayn L. Carey (who typed beyond the call of duty); Mitch Coddrington; M.

Conti; W. Custer; B. Davisson; S. De W.; K. Dexter; Duffeys; Patricia Eakins; J.E.; Ken Fields; GSF; J.F.; M&NF; RMF; C. Friedmann; B. Goldwater; E.G.; L.G.; E.H.; S. Hirth; M&M Helm; Duncan Holmes; K.H.; Bob Jeffries; Shirley J.; Sam Julty; Dick Kagan; E. Kashins; W. Langeweische; Bill Lear; P. Lert; B. Linda; B. Loader; Stormont Mancroft; J. Mayers; Henry McKay; S. & A. Martindale; Bill, Peter, & Virginia Miller; Roz Moore; A&E Munford; Lila Mukamal; A.M.; J.M.; M.M.; Chuck & Dorothy Neighbors; Fred N-S; J.N.; D.N.; Robbie N.; T.N.; Patricia O'Leary; Gail Olsen; Bill Pickup; R.P.; S.P.; L. Penta; Don Pemberton (of Cessna —many, many thanks); Regis Q. & family; Peggy Romeo; J.R.: Sharon R.; Lou Sandler & Famille; the Schneiders & Slotes; L&B Smith; N. Slepyan; L. Shapiro; Anne Saxon; Clara Schwabe; J. Sloan; A.S.; C.S.; D.S.; L.S.; P.S.; A.T.; J.T.; M.T.; Don Typond; Type Systems—a big hand; M. Vassallo; G. Van Holstein; R.W.; S.W.; D. Wingate; Elly Weiss; T.J. Wright.

Last but by no means least is Jim Neyland, without whom this book would never have come to pass. So, once more into the breach, read on—and please enjoy.

Introduction

At the risk of impertinence toward those several other writers in the field of aviation, there are not—when you come to think about it—so very many books within this sphere that could be described as "fully useful." Indeed, you can possibly number the invaluable ones on the fingers of your hands.

The reason for writing *Basic Guide to Flying* was to provide in reasonably compact form a ready reference for the beginner, for the low-time pilot, and for those pilots who, for reasons of budget, can't keep themselves as up-to-date as they would wish. We should not forget that the reason we fly in superjets today is not because some eccentric millionaire of long ago donated sums to a foundation to solve the problem of flight. We owe it largely to a couple of bicycle mechanics whose fingernails were dirty and whose pedigree dates back to some equally sweaty (and prehistoric) forebear, who very possibly had something to do with the invention of the wheel. So flying is not necessarily for the very rich, as J. P. Morgan once suggested yachting was. And it shouldn't be.

Part I of the book is for the beginner, the person who knows little or nothing of the intimacies of flight. Part II has to do with training. I've tried to concentrate on those areas that seem to cause the most confusion to the tyro and to provide information that seemed to me to be the most helpful. This section is intended to be used also for quick reference for those pilots who, like myself, occasionally forget whether latitude is actually longitude and vice versa. (It doesn't really matter too much because it's only nomenclature—as long as you know what you're doing!) But it can lead to problems if you're communicating with someone else, as in a Federal Aviation Agency (FAA) written exam, who assumes that when you say right you mean right and not left, which is what you may actually have meant.

Part III contains a chapter on careers. Many people like the idea of flying for a living; this is both an effective way of exercising one's responsibilities to one's fellow man as well as earning a living in an intriguing environment. A career in flying is for the man or woman who likes the idea of serving others while at the same time enjoying the challenge that a relatively alien environment, the skies, presents to man. This challenge is every bit as exacting as those mysteries of the seas that exercised our ancestors' imaginations.

For those who prefer to earn their daily bread in other ways but who are into flying as one of the more interesting facets of twentieth-century life, Part V contains a section of airplane ownership. It doesn't go into it in any great depth—it's more to stimulate the imagination. Pilot reports in miniature are also provided, together with the vital statistics of the airplanes mentioned. And there's a section on airplanes you can build yourself, which, again, is not exhaustive but which just might whet your appetite.

Much of the fun in designing an ideal airplane is in deciding with what it will be equipped. Although inertial navigation devices are still rather expensive for all save the military and the commercial airlines, there are numerous other devices that are not so expensive. So Part V contains a review of the equipment that conceivably might interest you, such as distance-measuring equipment: Dial a switch and the machine will read out how far you are away from a point. There is no need for your sextant any more; radio waves do the hard work for you.

Part VI includes a little history and a general overview of aviation as it existed up to the time the book was written. For further reading there's a fairly extensive bibliography.

As John Donne, the Elizabethan metaphysical poet, wrote, "No Manne is an Islande," and this is peculiarly true of that loose-knit fraternity of the skies. Several books on a particular subject by different writers can very quickly give a person a more thorough understanding of that subject than by simply relying on one source. Just as any writer has his hangups, so too do readers: What produces mental indigestion for one may satisfy another's appetite.

For current information *Air Facts*, *Air Progress*, and *Flying* are best known in the East. *Plane and Pilot* and *Private Pilot* are better known in the West. The Aircraft Owners & Pilots Association (AOPA) house magazine *Pilot* is also popular. Magazines for professional pilots include *Business and Commercial Aviation, Professional Pilot,* and *Flight Magazine. Aviation Week and Space Technology* has a section devoted to business aviation, and the Airline Pilots Association (ALPA) also produces a magazine for its members. Last but not least is *Sport Flying,* the house magazine of the Experimental Aircraft Association, whose members continue to bear the torch that was lighted December 17, 1903, by the brothers Wright.

PART I

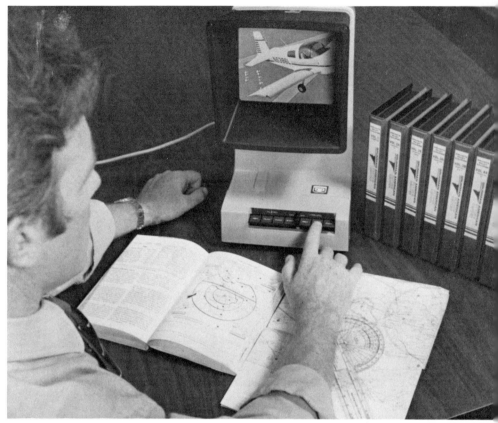

Training programs produced by aircraft manufacturers use modern techniques to lower cost and time in qualifying pilots. Film strips and cassettes provide individualized tuition, as seen here.

1

For the Beginner

A few years ago an airplane company ran a series of ads showing a mechanic in overalls smiling at the reader. The caption read: "If I can fly—you can fly." The copy then went on to extoll the pleasures of flight and how the reader could enjoy a demonstration flight for the attached coupon and payment of five dollars.

Actually, one of the best ways to begin to learn to fly is to take advantage of one of these five-dollar schemes, which are still offered from time to time. Take the coupon to the appropriate aircraft dealer nearest you, who will, provided the weather is reasonably clear, take you up for a short flight. You'll be sitting in the left seat and will have a chance to try your hands at the controls. When you come down, you should have some idea of whether you want to go any further. To get an even better demonstration, offer to pay for additional flight. It's worth it.

To learn to fly well enough to carry others aloft requires the expenditure of about a thousand dollars. If you're a sailboat or canoe enthusiast, a thousand dollars can go quite quickly in routine boating almost without your noticing it. If you whitewater kayak, or ski in winter at all, you won't express much surprise when you add up the cost of a season's sport—packed into four or five all-too-brief months. If golf or scuba-diving is your sport, a thousand dollars is not a lot. And if you motor race or rally, you'll know that one thousand is slight indeed when it puts into your hands an entirely new skill and opens a totally new dimension for you to command. The most important question is how to spend your money in acquiring this new skill as efficiently as possible.

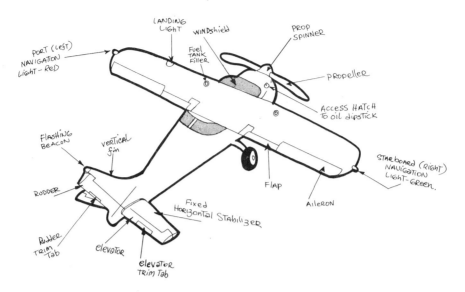

FIG. 1.1

You can be taught to fly in as little as two weeks, but that's pushing it. It is advisable to take a little longer to assimilate all the details. Not only do you have to learn to control a new machine in a new element, but there's quite a bit of book learning to be acquired. If you take longer—three to four weeks—you'll remember much more of what you've learned.

If you live in a large city, the chances are that there are several good training schools nearby. Modern teaching methods at their best are used in flight training these days. In the larger schools extensive use is made of simulators—ground-based equipment that partially duplicates the cockpit of an airplane—in which the student can get the mechanical feel of an airplane and simulated flight without leaving the ground. One great advantage in using simulators is that an instructor can stop the flight at any point, discuss the technique that needs improving, and then observe as the student tries it again. This can't be done in flight.

In addition to well-established schools, there are now training programs that have been set up by the aircraft manufacturers themselves. Cessna, the largest seller of general aviation aircraft in the world, has at its pilot-training centers reduced the number of hours its students need to acquire private pilot certificates to thirty-nine hours; the national average is about sixty hours.

Like the programs of Piper, Beech, and Grumman American, the Cessna training methods are based on carefully programmed and integrated instruction. The course is divided into stages of instruction, with the completion of one stage leading logically to the next. Ground-school work is combined with flying work so that a satisfactory balance is achieved. The theory of flight is taught along with the practice of flight so that boredom is less likely to set in.

This isn't necessarily the best way for everyone. There are fine instructors almost everywhere in the United States. If you live in the country, you could check around the neighborhood Fixed Base Operators (FBOs) to see if instruction is available. Often, smaller FBOs are able to charge lower rates than the chromium-plated flight-training establishments, which inevitably have larger overheads. The secret is to shop around to see where you're most likely to get the most for your money.

Once you've started, you can save more by buying block time on an aircraft. This means that you agree to pay in advance for a specified number of hours' use of a training airplane. Discounts

FIG. 1.2

What comes with the private pilot's package—a completely programmed training schedule that includes all the instruments and charts you'll need (with instructions on how to use them).

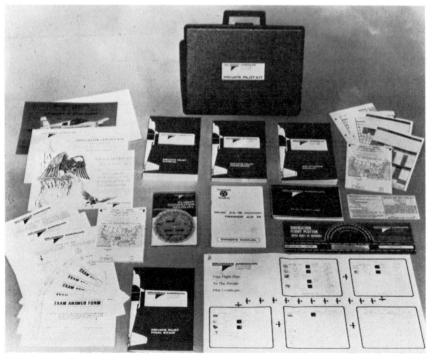

vary from place to place, but between 5 and 10 percent is normal. However, for your first two or three lessons, pay as you go. There are some instructors—don't let anyone tell you there aren't—who are only interested in building up sufficient hours to enable them to move on up the ladder to a better job. A sad fact is that in the past instructors have been regarded as necessary hacks, just cogs in the machine to make money for others. Not too much thought was given to the necessity of having good instructors if the country was to have good pilots.

The professionalism brought to the teaching field by the manufacturers and by such firms as Sanderson Films, Inc., Aero Products Research, and Jeppesen will accelerate the changes in the sort of instruction that is available. Indeed, improvements are already here.

PRELIMINARY STEPS

Before you actually start learning, you must have a medical examination. The FAA appoints doctors around the country to conduct these medicals. You will take what is called a third-class medical. The only medical problems that might disqualify you are such conditions as diabetes (which has to be controlled by a drug), epilepsy, alcoholism, and psychotic behavior. Vision, if it can be reasonably corrected and if you're not totally color blind, is no problem. Even amputees can pass this medical exam. However, it is advisable to get the details from your regional flight surgeon's office.

Your student license can be obtained once you've actually started. The other item you'll need is a Federal Communications Commission (FCC) certificate so that you can use a radio in the aircraft. Don't worry about either of these initially. (See Figure 1.3 for a reproduction of the necessary FCC forms.)

THE WALKAROUND

To take a closer look at an airplane, we'll choose a typical family four-seater, not too fast, with 130 mph as its cruising speed. (See Figure 1.4.) You can take your pick of high wing or low. For this example we'll use the high-winger because it is just a bit easier to get into and more of these planes are available.

FCC FORM 753-A
MAY 1965

FEDERAL COMMUNICATIONS COMMISSION
GETTYSBURG, PA. 17325

FORM APPROVED
BUDGET BUREAU NO. 52-R46.7

APPLICATION FOR RESTRICTED RADIOTELEPHONE OPERATOR PERMIT BY DECLARATION

A. USE TYPEWRITER OR PRINT IN INK. Signatures must be handwritten.
   Be sure to Complete all items including 7, 8, 9, and 10.
B. Enclose fee with application. DO NOT SEND CASH. Make check or money order payable to Federal Communications Commission.
   See Part 13, Volume I of FCC Rules.
C. No oral or written examination is required. Applicant must be at least 14 years of age.
D. U.S. NATIONALS AND U.S. CITIZENS: Submit one application to FCC, Gettysburg, Pa. 17325. U.S. Nationals who are not U.S. Citizens must attach a copy of certificate of identity.
E. ALIEN PILOTS: Submit one application to FCC, Washington, D.C. 20554. ALSO complete and attach FCC FORM 755.

DO NOT WRITE IN THIS BLOCK

1. REASON FOR APPLICATION
   NEW PERMIT
   NAME CHANGE (Attach Present Permit)
   REPLACE PRESENT PERMIT DUE TO ITS CONDITION (Attach Present Permit)
   ORIGINAL PERMIT IS LOST OR DESTROYED. IF FOUND I WILL RETURN IT TO FCC. A REASONABLE SEARCH HAS BEEN MADE FOR THE PERMIT
   OTHER (Specify)

2. NAME (Last) (First) (Initial)
   PERMANENT ADDRESS (No. & Street)
   (City) (State) (ZIP Code)

(Cut on this line)

W A R N I N G

MAKE SURE THE RADIO STATION HAS A CURRENTLY VALID STATION LICENSE.
THIS PERMIT DOES NOT AUTHORIZE ANY ADJUSTMENT OF THE TRANSMITTER THAT MAY AFFECT THE PROPER OPERATION OF THE STATION.
WATCH IF THE INTERNATIONAL RADIOTELEPHONE DISTRESS SIGNAL, DISTRESS MESSAGES HAVE ABSOLUTE PRIORITY.

PROHIBITIONS: USE OF OBSCENE, INDECENT OR PROFANE LANGUAGE.
UNAUTHORIZED DISCLOSURE OR USE OF MESSAGES.
SUPERFLUOUS, FALSE OR DECEPTIVE SIGNALS OR COMMUNICATIONS.
TRANSMISSION OF UNASSIGNED CALL SIGNALS.

OBSERVE REQUIREMENTS FOR TRANSMISSION OF STATION IDENTIFICATION.

THIS PERMIT MUST BE CONSPICUOUSLY POSTED AT THE PRINCIPAL OPERATING LOCATION OR KEPT IN YOUR PERSONAL POSSESSION, DEPENDING UPON THE RULES GOVERNING THE PARTICULAR CLASS OF STATION BEING OPERATED.

KNOW YOUR RADIO LAWS, TREATIES, RULES AND REGULATIONS. FAILURE TO OBSERVE THEM MAY LEAD TO SUSPENSION OF THIS PERMIT.

3. DATE OF BIRTH
   MONTH DAY YEAR
   PLACE OF BIRTH (City, State, Country)

4. ARE YOU A CITIZEN OF THE U.S.? (check one)
   BY BIRTH
   BY NATURALIZATION OF MYSELF UNDER CERTIFICATE OF NATURALIZATION NO.
   COURT OF ISSUANCE
   DATE ISSUED
   BY CERTIFICATE OF CITIZENSHIP NO.
   DATE ISSUED
   OTHER (If by none of the above three, check this box & explain)

5. Have you been convicted in the last ten years of any crime for which the penalty imposed was a fine of $500 or more, or an imprisonment of more than one year?   YES   NO

(IF YES, FURNISH DETAILS FOR EACH CONVICTION GIVING DATE, NATURE OF CRIME, COURT IN WHICH CONVICTED, NATURE OF SENTENCE AND WHERE SENTENCE WAS SERVED.)

6. DO YOU HAVE ANY OF THE PHYSICAL DEFECTS LISTED BELOW OR ANY OTHER DEFECT WHICH WILL IMPAIR OR HANDICAP YOU IN PROPERLY USING THE PERMIT FOR WHICH APPLYING? (Check below, if yes attach details)
   SPEECH IMPEDIMENT   YES   NO
   ACUTE DEAFNESS   YES   NO
   OTHER   YES   NO

APPLICANT'S CERTIFICATION

I certify that I am the above named applicant; that the facts stated in the foregoing application and all exhibits attached thereto, are true of my own knowledge; that I have need for the permit herein applied for; that I can transmit and receive spoken messages in English; that I can keep at least a rough written station log in English or in some other language in general use that can be readily translated into English; that I am familiar with the provisions of treaties, laws, rules and regulations governing the authority granted under the permit herein applied for; that I understand that it is my responsibility to keep myself currently familiar with such provisions; that I will preserve the secrecy of radio communications as required by law and that I will faithfully adhere to any requirements of law at all times, that this obligation is taken freely, without mental reservation or purpose of evasion; and that I will and faithfully discharge the duties of the office obtained through my employment under this Permit if granted.

(Signature) (Date)

WILLFULL FALSE STATEMENTS MADE ON THIS FORM ARE PUNISHABLE BY FINE AND IMPRISONMENT, U.S. CODE, TITLE 18, SECTION 1001.

FEDERAL COMMUNICATIONS COMMISSION
GETTYSBURG, PA. 17325

OFFICIAL BUSINESS

Penalty for private use, $300

FCC FORM 753-B
MAY 1965

UNITED STATES OF AMERICA
FEDERAL COMMUNICATIONS COMMISSION

RESTRICTED RADIOTELEPHONE OPERATOR PERMIT

This PERMIT, when countersigned by the Secretary of FCC, authorizes

to operate licensed radio stations for which this
is valid under Rules and Regulations of the Commission and for the lifetime of the holder subject to suspension pursuant to the provisions of Sections 303(m)(1) of the Communications Act and the Commission's Rules and Regulations. This Permit issued in conformity with Paragraph 90), International Radio Regulations, Geneva 1959.

PERMITEE SIGNATURE

PRINT OR TYPE FULL NAME

SIGNATURE

NAME AND MAILING ADDRESS

FIG. 1.3
Application for restricted radiotelephone operator (official form of the FCC)

FIG. 1.4

Cessna's Skyhawk 172 series (1973 model)

Walking toward it, you will notice that the propeller, which is attached to the engine, has two blades, and there's a sharply pointed cone at the center. Beneath the engine cowling there is usually a shock-absorbing strut and wheel—the nosewheel.

If you open a little hatch on top of the engine cowling, you can look inside and see an engine that appears to be quite small, compared to that of a car. A closer look reveals a dipstick—with "oil" written on it—which, unlike a car, must be unscrewed before you inspect the oil level. Right next to the dipstick on this airplane is a square-shaped knob labeled "drain." If you are curious to see what it does, take care where your feet are, for when you pull this knob, fuel will pour out onto the ground.

The purpose of the dipstick is obvious—to check that the engine has sufficient oil. But what about the knob? It is, as the label indicates, a drain, and its purpose is to make sure that no water gets into the gas. Car engines don't like water mixed in their diet of gasoline and air, and aircraft engines are even fussier. They rebel, they stop working, and the pilot must then glide to earth for a landing.

To be sure this doesn't happen, fuel systems in airplanes are designed so that any water (which is heavier than gasoline) in the fuel lines is led to a low point where it can be easily drained without losing much fuel. Because water freezes before gasoline, an accumulation within the system could conceivably reduce or even stop the flow of fuel to the engine. And if the weather is warm, water could easily make its way to the carburetor and so to the engine. The conscientious pilot makes a check of his fuel system by draining some fuel from the line into a clear container and checking it visually for water before setting out on a journey. If there is any water in this test sample, he doesn't fly until all the water has been removed from the fuel. He checks his gas and oil, and then makes his walkaround.

The walkaround is important. It's a last chance to inspect the exterior of the airplane before leaving the ground. Some pilots are rather casual about it. They know their airplanes well and, we're told, have some sort of psychic rapport that is activated by a kick of the tires before takeoff. Personally, I'm not a very courageous person, therefore, I tend to be slightly compulsive about a thorough walkaround before venturing into the air.

Inside the cabin we make sure that the master switch is off and that the mixture control, if any, is in the cut-off position—a double safety precaution. Every pilot has his own starting point for his walkaround, and most instructors will suggest to the beginner on his very first walkaround that he choose the point at which he wants to start—and then stay with it. I find the propeller a very logical place to begin.

The propeller is actually an airfoil; when it rotates it generates lift that pulls the airplane through the air. On helicopters the propellers, or rotors, are placed so that in addition to providing lift they also provide forward and backward motion.

Unfortunately, because they swing quite close to the ground, propellers get small scratches and nicks from stones, mud, and other debris. Even the smallest scar will put a propeller slightly out of balance. So before we take off, we must make sure that the prop

The Walkaround.

1. Open cockpit, check switches OFF. Remove control locking device.
2. Check flaps and ailerons secure, and ailerons move freely.
3. Check wingtip secure.
4. Check fuel drain for water in fuel and check filler cap secure. Check mainwheel, brakes, etc. Check oil.
5. Propeller: check for damage and security. Check nosewheel.
6. Same as 4, plus pitot tube, minus oil.

7. Same as 3.
8. Same as 2.
9. Check static source.
10. Check rudder and elevator assembly. Check underneath fuselage for oil slicks.
11. Check static source.
12. If you're satisfied all is well, get in, close and lock the door and fasten your seat belt.

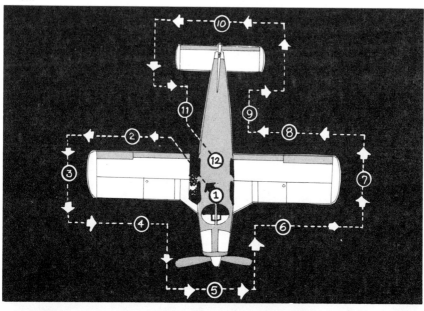

is in good shape. If there's a nick, we might ask a mechanic to come over and examine it. Simple things can be "dressed" with a round file, thereby preventing any out-of-balance forces. If not taken care of, a damaged prop blade could fly to pieces in all directions.

While we're at the front of the airplane, we check the forward cowl openings, which provide cooling air for the engine. It is important to see that there are no birds' nests or fieldmice or other creatures inside. Look at the nosewheel for scratches, abrasions, and pressure, and check the nose strut to see that it still has its shock-absorbing qualities, with no leaks of hydraulic fluid from the oleo strut.

Moving around the airplane's right side, climb on the wing strut to check the fuel level in the wing tank. Fuel gauges get better and better each year, but the only way you can be sure that there is enough fuel is to see it for yourself. Also, you have to put the gas tank caps on yourself to know they won't come adrift in flight. If they did, the airflow over the wing could quickly syphon out a full tank of gas.

The front edge of the wing is important. It is this that meets the oncoming air and causes it to flow over and under the wing. At a certain speed, when the edge of this wing is presented to the air at an appropriate angle—the angle of attack—the air moving over the top of the wing creates a low-pressure area, thus generating lift.

If for some reason the leading edge is deformed—by ice, perhaps, or simply by dents—it affects the airflow over the wing. It is for this reason that larger aircraft are fitted with expensive heating devices within their leading edge surfaces to prevent the formation of ice.

We now move outward to check the wing tips. On most modern aircraft these are not an integral part of the wing but are attachments to help the air flow smoothly off the wing and to reduce wing-tip turbulence and drag.

At the back edge of the wing we note a section that, while it appears to be a part of the wing, has the potential to move on its own, up and down. If we move it up while looking at the other wing, we shall see its twin moving down. If we reverse the process, the other reverses as well.

These sections are called ailerons. When one side goes up and the other goes down, it interferes with the flow of the air over and under the wings, causing the airplane to bank and to turn. We'll examine them and the way they work more closely later. For the time being let's just make sure that they are securely fastened to the wing and that the actuating rod—which you can see when you're moving the aileron up—is secure.

Just inboard of the ailerons is another section—the flap. The flap is used to increase both lift and drag and in some airplanes is used to provide additional lift for takeoff. Most aircraft with flaps use them to provide drag with lift for slower approaches to landing. On this airplane the flap mechanism is operated by an electric motor in the wing, so we will only visually check that it is secure. We will run a full check of flap operation from the cockpit.

While we're here, we'll check the brake hydraulic line for any leakage and the disc brake pad to make sure it hasn't worn down too much. We can check the tire, too.

Just forward of the wing strut on the right side, in the side of the airplane, is a circular attachment with a needle-sized hole on the paneling forward of the door. This is one of the two sources of what is called static air, which some of the instruments use. For the moment just see that the hole at its center isn't clogged in any way.

The fuselage aft of the cabin is quickly checked. This is really a long tube that acts as a balance to the front half of the airplane and serves to carry the controls to the tail area. Inside, behind the luggage compartment, we find the battery, a solenoid, plus a couple of fuses. The tail unit consists of a horizontal stabilizer and elevator and a fin and rudder in the vertical plane.

The rudder, unlike the rudder in a boat, is not used to turn the aircraft at all. It is merely there to make the back end of the airplane follow the front. If you've ever driven a car along an icy road, you may have found that the back end sometimes does not want to follow the front, which you correct by steering toward it with the front wheels. An airplane works in much the same way, except that it can, in effect, skid in two directions. It can skid like a car, outward from the main axis, but it can also skid inward, which a car can't really do. In the air this is called a slip. The rudder is there to help the pilot balance his turns so that he neither skids nor slips when he's turning.

If the rudder is for balancing the airplane when it turns, the horizontal tail at the rear of the airplane is there to maintain the balance of the fuselage. This may sound a bit complicated but it really isn't. At the front end of the airplane is the engine and behind it the wings and cockpit. To balance these, we need a lever. The fuselage of the airplane acts as the lever, while the horizontal stabilizer—the name for the fixed part of the little wings at the rear—provides a counterpoint for the weight at the front end. When we want to move the front end of the airplane out of the horizontal plane, we use a blending of power in conjunction with the movable part of the stabilizer. We don't actually need to use it

11

because adding or subtracting power—provided the airplane is trimmed or leveled off in flight—would achieve the same result. Use of the stabilizer speeds the result.

The moving part of the stabilizer is called an elevator, except in those airplanes that are said to have "flying tails," which means that the whole stabilizer moves. If you think that this movable surface is what you use to control the up and down movement of the airplane, forget it. What the elevator actually does is to control the angle of attack of the leading edge of the wing.

The throttle controls up and down movement. As we progress, you will see how you can fly this airplane at 60 mph in level flight and at 130 mph in level flight. At the slow speed, the nose will be pointing up much higher in the air because our angle of attack will necessarily be greater at that speed to generate additional lift. We'll achieve that particular angle of attack by easing back on the control wheel, which in turn will move the elevator. At the 130-mph speed we employ just the tiniest amount of forward pressure so that although we are actually flying straight and level, you might imagine that we are pointed down very slightly.

For the time being we simply make sure that the elevator is fastened to the stabilizer and that the device that looks like a mini-aileron on the trailing edge of it (it's actually a trim tab) moves gently in the opposite direction to the elevator when we move the elevator up and down. While we're at the end of the airplane, we can check to see that the rudder cables are secure and that the underside of the fuselage isn't showing any oil streaks—this is frequently the first place an oil leak will show up.

We now walk forward. Behind the rear cabin window the antenna of the Emergency Locator Transmitter (ELT) waves like a gossamer thread in the wind. We make the same check on the left side of the aircraft as the right, but there are two additional items to note. The first is a small hole at approximately the center of the leading edge of the left wing—the stall-warning device. When the air flow starts to break down over this surface, as it will when nearing the point of stalling, it causes a small reed to warble in the cockpit, warning us that we may be about to stall the airplane. It is not very efficient; a more useful device, which is becoming increasingly popular, is an angle-of-attack indicator. This is an electrically operated device mounted on the wing, with an instrument in the cockpit showing constantly at what angle of attack the airplane is being flown. This aircraft doesn't have an angle-of-attack indicator, so we shall make sure the stall-warner isn't clogged up.

The other item we must inspect is a tube suspended from the wing. It has a small hole in it, facing forward, into which the

Slung below the wing, the pitot tube measures the pressure of ram air and by differentiating this with air from the static source, provides a readout of airspeed on the airspeed indicator (ASI).

oncoming air, of the aircraft in flight, is forced. This is called the Pitot tube; it feeds the oncoming air to the airspeed indicator, which, balancing this against the "static" air (remember the little holes on the right and left sides?), gives the airplane's airspeed in miles per hour.

Most aircraft provides electrical heating for the Pitot tube so that it doesn't get iced up. If the Pitot tube should get clogged—even a small insect can do it—you'll suddenly become aware that you've no airspeed. A working Pitot tube is important to the pilot because by knowing that we have sufficient airspeed and by maintaining it we are able to stay aloft.

One final check—the other static port and the wing tank—and with the wheel chocks and the tie downs removed, we can start to think about flying. Let's get on board.

INSIDE THE AIRPLANE

Let's have a look at the cabin. There are locks on both the right-hand and left-hand doors. Opening the left-hand door (the side on which the pilot in command usually sits), we see what looks like a decapitated steering wheel sticking out from the dashboard—the control wheel. There may be another control wheel facing the right-hand front seat. The dashboard contains numerous dials, the meaning and use of which we'll learn later. An airplane's dashboard is called an instrument panel. (See Figure 1.5.)

13

FIG. 1.5

The instrument panel. Situated directly in view of the pilot in command are the main instruments used to control the aircraft. At center, the artificial horizon (A/H), or attitude indicator. Immediately beneath it, the directional gyro (DG), which is set to the magnetic compass. To the left of the A/H is the airspeed indicator (ASI); and to the right, the altimeter, which tells the pilot how high up he is. To the left of the DG is the turn-and-bank (a turn coordinator is found more and more on newer aircraft); and to the right, the vertical speed indicator (VSI), which tells the pilot at what rate—in feet per minute—he is climbing or descending.

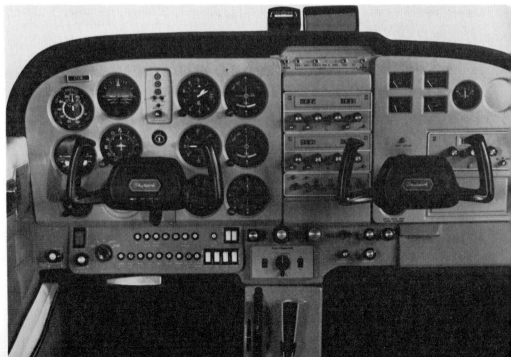

A piece of metal with "control lock" printed on it has been pinned through the arm of the control wheel. We took this out when checking the master switch and mixture control before the walkaround. We now check that all the electrical circuits are switched off. We are going to fly, but before we start we must check out the instrument panel switches. The red switch, split in two, with the words "Master Switch" below, gives us our clue. The down position means that the circuits are off; the up position, that they are on. Ours is in the down position, so the current must be off.

One other switch that must be checked is the ignition switch.

We are careful about this step because even though the switch may be off, it is possible—though not very likely—that the prop could burst into life. And for this reason we always handle the prop as if it would.

Now, settle into your seat. Feel around in the middle of the seat, below the cushion level, and you'll find a small bar. If you move it, you can slide the seat back so that you can get in comfortably. Don't worry about the charts for the moment—I'll look after the navigation today. There's a lap strap and a shoulder harness that fits into the buckle of the lap strap. Don't do it up tight just yet. First, slide the seat forward again until your feet are comfortably resting on the pedals. Now you can tighten the belts.

Before we start up air, we shall preflight the aircraft from a checklist. But first, let's go over the instrument panel and see just what those dials mean. At the center of the panel in front of you there's a large dial with lines and with what vaguely resembles the outline of an airplane. This is called an artificial horizon (A/H). It tells you about the airplane's attitude—that is, whether the airplane has its nose up or down and whether its wings are level or banked. A gyroscope within the A/H provides the information shown on the dial. Because there are a number of instruments that use gyroscopes, we may as well examine one.

A gyroscope looks like an ancient philosopher's master plan of the universe. Its center is a spinning wheel mounted to a single axis. This axis is universally mounted so that only the center of gravity is fixed; the wheel itself is able to turn in any direction around this point. If you were to build a gyroscope (Figure 1.6), it would go something like the following.

First, take a solid wheel—a flywheel, for example—and mount it on an axle through its center. The wheel—if you want to make the sort of gyro that you could use in an airplane—should have little cups around its outer edge into which you can squirt air. Now put

FIG. 1.6

A gyroscope

the flywheel and its axle within a circular supporting ring, within which they may spin. Next add a further supporting ring, the bearings of which are at a 90-degree angle to the flywheel's axle and in the same plane as the flywheel itself. Finally, make a supporting frame for that outer ring that will support the structure in a horizontal plane. You now have a working model of a gyroscope. To make the wheel spin, direct a jet of air into the wheel. If you don't have an air hose, wind some string around the wheel and give it a hearty tug to set the wheel in motion.

Once the wheel is spinning, you can point the center axle of the gyroscope in any direction without altering its geometrical center. The spinning of the wheel brings the gyroscope to life; it acquires a high degree of rigidity. No matter how much you turn the base of it about, the center axle still wants to remain in the same place. However, gyroscopes do have one bad habit—precession. This means that if you apply a force around its horizontal plane or axis, instead of turning the gyro will precess or move in the vertical axis or plane 90 degrees to the direction of the applied force.

The more solid the flywheel of a gyroscope, the more resistant it is to being disturbed. And if, for reasons of weight, you don't have a very solid flywheel, you can make do by spinning it faster. (Airplane gyros commonly spin at more than 20,000 rpm.) Gyroscopes provide information about attitude (as in the A/H), direction, and turns. The gyroscope presents this information to the pilot in the following way: When the wings of the mini-airplane within the A/H are level, the actual airplane's wings are level. If the dot that represents the airplane's nose and the wings are above the artificial horizon line on the instrument, then the real airplane is climbing. If you examine the A/H very carefully (Figure 1.5), you will see that we could measure the amount of climb (or descent) in terms of the actual bar width. And if the dot and the wings are below, then the real airplane is descending. For pilots flying on instruments, the A/H provides primary information about the airplane's attitude to the horizon.

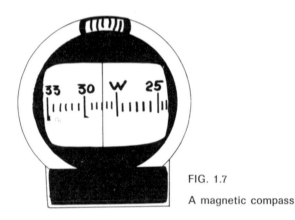

FIG. 1.7

A magnetic compass

Just beneath the A/H is a compass card. This is also operated by a gyro, and by setting it to the aircraft's compass, one can avoid those errors that plague the magnetic compass. (See Figure 1.7.) The directional gyro (DG) is subject to that gyroscopic error called precession. So the wise pilot checks his DG every ten minutes or so against the wet or magnetic compass.

To the left of the A/H is the airspeed indicator. As already mentioned, this tells you how fast the airplane is traveling in relation to the air it is flying in. It does not give ground speed. Air speed is vital to the pilot: If he has sufficient speed to provide lift, he can fly, if he hasn't, he must come down. Airspeed is used to tell us about our lift coefficient when we are taking off or landing. An angle-of-attack indicator (Figure 1.8) would be much more useful, but most manufacturers don't seem to have heard of them. Airspeed is also useful for telling the rate at which you are traveling through the air, and is used in computing whether you have a headwind or a tailwind, and so on. (See Figure 1.5.)

The point is that if the lift coefficient is such that the wing cannot support the weight of the airplane, the aircraft is designed to stall—that is, to try to find a way in which lift may be redeveloped. Lift coefficient is reduced when our angle of attack is too great. To avoid stalling we must try to keep the angle of attack within bounds to maximize lift. (Later you will learn in more detail how to stall an aircraft so that you'll be familiar with the condition and can avoid it except in practice.)

To the right of the A/H is the altimeter. (See Figure 1.5.) This is simply an ordinary aneroid barometer that has been set to read altitude rather than air pressure. Because the air gets thinner the higher you go, the barometer measures this decrease in pressure in feet.

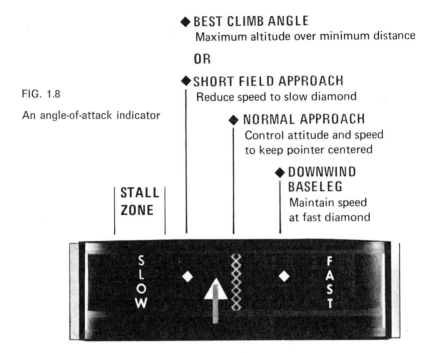

◆ BEST CLIMB ANGLE
Maximum altitude over minimum distance

OR

◆ SHORT FIELD APPROACH
Reduce speed to slow diamond

◆ NORMAL APPROACH
Control attitude and speed
to keep pointer centered

◆ DOWNWIND
BASELEG
Maintain speed
at fast diamond

STALL
ZONE

FIG. 1.8

An angle-of-attack indicator

Immediately below the altimeter is the vertical-speed indicator. This tells you at what rate you are climbing or descending. It's a little bit slow in presenting this information, because it has a certain inertia to overcome, but it will tell you instantly in which direction you are moving if a change should occur. Most pilots like to climb and descend at about 500 feet per minute (fpm) because this is easier on the ears in adjusting to changes of air pressure. It is also a useful rate to use when you're flying by instrument rather than by visual contact.

To the left of the DG (Figure 1.5) is the turn indicator. The turn-and-bank, as the device used to be called, is also a gyro instrument. In the old days, before the A/H was invented, it was the only way the pilot had of knowing that his wings were level and that he was not slipping or skidding. Old-time instrument pilots used to fly routinely through clouds using merely this instrument, a magnetic compass, and an old-fashioned altimeter.

The top part of the turn indicator measures the angle of bank in a turn; the bottom part—the ball part—is like a builder's level, measuring any unbalance in the actual turn. (See Figure 1.5.) If the rate of turn is too fast for the angle of bank, the centrifugal force from the turn will move the ball toward the outside of the turn. If the rate of turn is too slow for the angle of bank, the

18

absence of centrifugal force will move the ball toward the inside of the turn. As you start your lessons, your instructor will probably tell you to "Step on the ball." He means that you should use left or right rudder to offset whichever way the ball is moving.

Next is the tachometer. (See Figure 1.9.) If you've ever driven sports cars, you'll remember that a tachometer is used for optimizing the power produced by the engine when shifting gears. An airplane's tachometer measures the speed at which the engine is turning in revolutions per minute (the engine in almost all aircraft is directly linked to the prop). In more complex airplanes you will find a different method used to measure the prop speed.

For climbout we will be using full power, and when we get to our cruise altitude, we'll bring the throttle back. Just how much we shall throttle back will be indicated on the tachometer. Today we'll use about 2,500 rpm, which will give us not only good range but also a useful cruise speed.

There's another cluster of instruments relating to the engine, but perhaps the most important at present is the oil-pressure gauge, which confirms that the engine is receiving the proper pressure of lubricating oil. Beside it is the oil-temperature gauge, which confirms that the oil is also doing its ancillary job, which is to help cool the engine. Then there are the fuel gauges. Earlier we checked the fuel level visually. We know—from the owner's manual—that this engine, with this fuel-tank capacity and at the rpm setting we will be using, can keep us up for a little more than four hours. We'll make a note of the time we get off the ground, but we shall also be able to make a cross-check of the amount of fuel we burn by watching our fuel gauge.

The throttle is the black knob in the center of the panel. The red knob with the serrated edge is the mixture control we noted was

FIG. 1.9

A tachometer

fully out before the walkaround. Some airplanes—like the Cherokee—have a pedestal on which the throttle and mixture control are mounted in a manner similar to twin-engine aircraft. The mixture control is used to measure the mixture of gas and air fed to the engine, as the gas must be mixed with a quantity of air (for oxygen) before the engine can burn it. To the left of the throttle is a little knob marked "Carb Heat." This provides warm air to the carburetor when we are running the engine slowly, which prevents the formation of ice in the engine. Unfortunately, gas, being very volatile, tends to cool rapidly when it evaporates, and when it passes through the carburetor, it can cause the water vapor in the air to condense into ice. That blocks the flow of gas and air into the engine, causing it to cough, sputter, and finally stop working if the situation isn't corrected. So for this engine we shall use carb heat whenever we slow it down, just to be on the safe side. (See Figure 1.10.)

Notice the vacuum gauge—that little dial in the center—and the ammeter. These instruments confirm that we have adequate vacuum pressure and electricity for our instruments. We won't go into that right now, but we want to keep an eye open for any alteration in the rate because it could indicate an incipient instrument failure or other problems.

FIG. 1.10

Carb Heat provides an answer to icing in the engine by feeding a flow of warm air directly to the inlet orifice of the carburetor, a favorite place for ice to form when atmospheric conditions enhance this possibility.

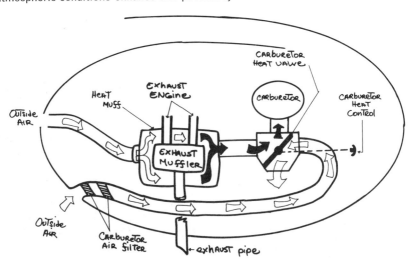

FIG. 1.11

A typical automatic direction finder (ADF)

Then there's the radio gear. I'll be looking after it for this flight, so I'll explain only that it is composed of a communications radio, which we use for talking, and another one for navigation. The latter also has a talk feature, which we sometimes use, but mostly it's for navigating. That needle in the dial belongs to the automatic direction finder (ADF), which is a nice gadget (Figure 1.11). The needle points to a radio beacon whose frequency you have dialed into the set. The other dial, the Very High Frequency Omnirange (VOR) receiver (belonging to the navigating radio), will show us what our magnetic bearing to the VOR is and, if we center the needle, will lead us to it. We call the bearings from the VOR "radials." As a navigation device the VOR and its receiver are more sophisticated than the ADF and radio beacon.

Now you can put the ignition key in. As you do, notice that there are detents marked "L" and "R" before you get to "Both." These detents are for the left and right magnetos. Twin magnetos are used as a further measure of safety; if the engine is already running, they can produce the electric current for the spark needed for ignition, even if the other electrical systems fail. When we run our engine check prior to taking off, we shall check the engine rpm drop when we switch from one magneto to the other. A certain drop is allowed, but it is monitored closely, as a large drop can indicate a possible malfunction. Finally, make note of the clock. Time and tide wait for no man, and the wise pilot always makes sure that he's back on the ground before his fuel runs out. Running out of fuel in the air can ruin your whole day.

You're the captain, so I'm going to call out our checklist for you.

1. Seats, seat belts, and shoulder harnesses: checked and secure?
2. Fuel tank selector valve: select BOTH.
3. Brakes: test and set firm. (You can use the handbrake while we start.)
4. Both radios and all electrical equipment: OFF. (We need all the power the battery can provide to turn the engine over, and the surge of power once it starts can cause radio equipment to malfunction.)

All right, now comes the drill for starting the engine.

1. Mixture: RICH. (Push the red knob in.)
2. Carburetor heat: COLD. (Make sure the little square-headed knob is pushed in.)
3. Primer: two strokes. (That should be enough if the day isn't very cold. Were the engine warm, we would skip this procedure.)
4. Throttle: open about one-eighth of an inch. (A quick way to gauge this is to pull the throttle all the way back, place your right forefinger up to the throttle friction nut—that's the serrated ring up by the panel—and then take it back to what you judge to be an eighth of an inch.)
5. Master Switch: ON. (Now you can push that double red switch up!)
6. Propeller Area: clear. (Yell "Clear!" at the top of your voice.)
7. Ignition: START. (Immediately release once the engine fires.)
8. Oil pressure: check. (If within thirty seconds of firing the engine you don't have an oil-pressure reading, switch everything off and call the mechanic.)

The engine is running sweetly and our oil pressure is in the green. While we warm up the engine, I'm going to call the tower, a very simple procedure. It goes something like the following:

1. Tune in the correct frequency of the person you want to call.
2. Call him by name.
3. Tell him who you are.

Having established contact, you next do as follows:

4. Tell him where you are.
5. Tell him what you want to do.

Here's an example of what it's going to sound like. We're at Washington Town Municipal, and this is what we're saying:

"Washington Ground, this is Goshawk Two Zero Four Zero Zero . . ."
"Goshawk Four Zero Zero Washington Ground."

"Washington Ground 400 is requesting permission to leave the ramp for the active, local VFR. Four Zero Zero . . ."

"Roger. Proceed to runway Two Three via taxiways Alpha and Bravo, up runway Thirty, then left via taxiway Delta. 4,000 feet available. Winds, light and variable. Altimeter, 30.05."

"Altimeter 30.05 Four Zero Zero."

FIRST FLIGHT

As we move off, we set the altimeter at 30.05. Taxiing an airplane is really quite easy. As you simply steer it by using the rudder pedals, you can forget about the control wheel. You might also note that pressing down on the upper tips of the pedals activates the brakes. You can use this for turning in tight quarters. While we're taxiing to the runway, I'm going to check the flaps. As there's virtually no wind, you can check the controls for free and easy movement. Just turn the control wheel left and right and visually check that the ailerons move properly. Now push the control wheel forward and back—if you look behind you out of the rear window you should see the elevator moving up and down.

Near the active runway there's a runup area. We turn the airplane's nose into the wind and now perform our pretakeoff checks. I'm going to call it off:

1. Parking brake: set.
2. Flight controls: checked for free and easy movement.
3. Fuel selector valve: BOTH.
4. Elevator trim wheel: *set for takeoff.* (You'll notice that the wheel set in the panel that falls beneath your right hand has a little white line on it. Set this to the position that reads TAKEOFF.)
5. Engine check: push the throttle forward until the tachometer reads 1700 rpm.
   a. Vacuum: check (for 4.6 to 5.4 inches.)
   b. Magnetos: check. Turn the key to the left detent, then note the rpm drop by turning key back to BOTH. Same procedure at right detent. Drop should not be more than 125 rpm with this engine. If it is, we'll turn around and go back.
   c. Carb heat: check operation. With warm air selected, the rpm will show a small drop of about 50 rpm if it is working properly. Bring it back to cold.

You can now throttle back to normal idle, about 1000 rpm.

6. Set flight instruments. We've already set the altimeter, but we should set the DG. We verify the runway direction against our magnetic compass; they seem to agree, so we set the DG to the compass. If the

compass did not agree with the runway heading—and we were certain we were at the right runway—we would make a note of the compass deviation and set in the runway heading. We'd also get the compass thoroughly checked before making any but a local flight.

7. Set radios. We set in the tower frequency and the frequency of a radio beacon toward which we will be flying, having identified it from its three letter code in Morse.

8. Final check that all doors, belts, seats, and so forth are all secure. *No smoking!*

I've changed the radio frequency from ground control to the tower, and I say, "Washington Tower, Goshawk Two Zero Four Zero Zero, ready for takeoff." Moments later, with little traffic around, the tower replies, "Four Zero Zero, cleared for takeoff."

We taxi onto the runway and line up the nosewheel with the runway's center line. We make a last check: that the DG, wet compass, and runway heading match. All you have to do is push the throttle smoothly all the way forward, and keep the aircraft going straight down the runway. The torque from the propeller will give the airplane just a nudge to the left, but a little pressure on the right rudder pedal will take care of that. When the airspeed indicator shows 60 mph, ease very slightly back on the wheel, and the airplane will start to leave the ground almost immediately. We'll climb out at 85 mph.

Throttle forward. Touch of rudder. We're coming back on the wheel now, and suddenly we're airborne. Keep an eye on your speed—85 mph—and keep a good look out for other traffic.

Perhaps the first thing you'll notice is that, unlike a car, the control pressures needed to make an airplane do what you want it to do are very light indeed. Then, as we make some turns, you may notice a similarity with the way one turns on a bike or motor bike. You actually bank to make the turn. Once we've reached cruise altitude we throttle back and start a little crosscountry flight. It's not too early for you to start recognizing landmarks on the ground, so take a moment to look out of the window; I'll take the controls. With the engine throttled back (and in noisier airplanes, with our earplugs firmly in place) it really is very peaceful. Note the altitude we're flying at—4,500 feet. That's because we are flying in a westerly direction. If we were flying east we would be at either 3,500 or 5,500 feet. We'll explain this point later.

Experiment with the controls. Try turning left—just turn the wheel very slightly. Notice how once you've started the turn, the aircraft wants to steepen unless you neutralize the wheel. Try turning to the right this time and watch the ball in the turn coordinator.

See how it moves slightly off center? That's because you need a touch of rudder in starting the turn. Try it again.

Landing is the one skill in flying that students have difficulty in mastering. First, slow the airplane down. A good landing starts early, which means that you set the airplane's speed ahead of time for the speed you want on final approach. There are several items in the prelanding checklist, so let's get them out of the way.

1. Mixture: RICH.
2. Power: set at 1750 rpm, which will give us a rate of 500 fpm descent the way we've trimmed the airplane.
3. We've applied CARB HEAT to avoid icing.
4. Fuel selector valve: BOTH.
5. Wing flaps: forget this for the time being. You can learn about flaps later.
6. Airspeed: 80 mph.

This is a little country airport and there's no control tower, but I have checked the chart and notice it has a Unicom frequency. Unicom without a control tower is an advisory station—usually run by the local FBO—on a frequency of 122.8. Most pilots call Unicom asking for the active runway when they're five to ten miles from the field—the result is an overload on the frequency most of the time in populated areas. Where there is a control tower, the frequency used is 123.0, and normally you use it to ask the gas truck to standby to fill up your tanks. Unicom is useful for informing other traffic in the area where you are and what you are doing. We listen to see if anyone's around we haven't spotted and then we'll call and tell them we'll be entering downwind for their runway 24.

Now I'll take the controls, but I want you to follow me through. Check very carefully again for any other traffic in the area—we have to keep a very close watch at uncontrolled airports, and we'll keep listening very carefully to that Unicom in case any other airplane announces itself. We finish the downwind, and as we start to turn base leg, we call to say that we are doing just that. Our base leg is a little tight, so we'll start our turn on final—with a last look to clear for traffic—a little early. Now we're lined up for our landing.

Now notice the perspective and try to remember how it looks. Notice how the runway appears, how it fits into the field of vision as part of an overall picture? Notice that even though we're getting closer, the perspective doesn't change. If it did, that would indicate that we were either climbing or descending from our glidepath.

Try to remember how it looks—it will help you when you come to do it yourself. And also note how I'm keeping our rate of descent very steady and our speed constant.

Now we're approaching the ground. I start to break our glide very smoothly, easing back the control wheel. The airplane begins to sink a little, and I ease the control wheel back a little more. Our descent steepens but slows, and I ease back more, keeping the airplane off the ground. Our speed is slowing, but we're not yet to stalling. Just a fraction more back pressure on the control column and she touches. As we roll down the runway, I let the nosewheel come down and apply gentle braking.

PART II

Solo: Suddenly everything's falling in the slot, and you've soloed.

A multi-tube assembly supports the engine at the front of the aircraft, which in turn attaches to the main fuselage at the firewall. Note the nosewheel attachment to this section. It is because the nosewheel is a part of the engine support structure that it is inadvisable to land too heavily with the nosewheel, since the shock could damage the tubular bracing here.

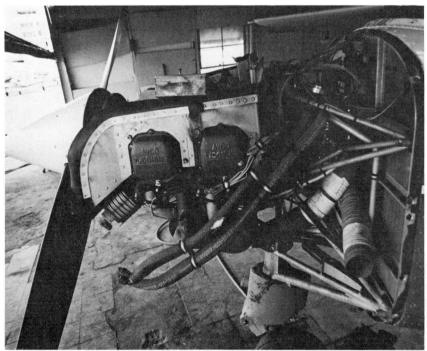

2

Preparing for the
Private License

Of all the motor skills you will learn in flying, those applied during landing are the most difficult to acquire. But once acquired, they are fortunately the most difficult to lose. It's a bit like riding a bicycle. When you start, you keep falling over, first to one side, then to the other. Then suddenly one day you can do it. Landing an airplane is not quite as easy as that, but as soon as you consciously perceive what you are doing right, you'll be able to retain it. Good landings start way back in the pattern, when you have the airplane under your control.

Ground-school materials and flying instructions are integrated in modern flight training leading to the private license; although you can't fly a book, a book can be a way of getting on the right track. We'll take a look at the actual learning cycle, including the sort of math and physics you have to know—don't worry, this involves common sense more than anything else. But first it may be helpful to chart out a typical flight-training program.

A FLIGHT-TRAINING PROGRAM

Familiarization with training airplane. This includes being responsible for the walkaround every time you fly and a growing understanding of how the various parts work.

Emergency drills. Fortunately, emergencies are very rare, but you'll want to know what has to be done if one should arise. The two principal areas of concern at the beginning are fires, in the cabin or fuselage or in the engine, and engine failure, which, as you will see, is treated similarly to an engine fire.

The walkaround plus. Increasing and deepening our knowledge of the reasons behind what we do, we learn more about cockpit checks, engine-starting procedure, and switch off after landing.

Now we fly. We learn the effects of each of the controls and get used to the light pressures that are involved. There's a certain amount of work to be done in ground school at this point, but it's mostly common sense. We learn about cruise power and how to use trim and take a first look at navigation.

Taxiing. By an admixture of ground school and flying, we learn about the importance of taxiing safely on the ground. Taxiing is fun, but taxiing an airplane in a wind is not the easiest of tasks— there's a right way and a wrong way. At this point we may also make some takeoffs from turf strips or make landings on grass surfaces.

Straight and level flight. In flight once more we begin to learn how to achieve straight and level flight—and believe it or not, the moment you look as if you are doing it gracefully, your instructor will start you on the next step: climbing and descending turns.

Climbing and descending. We're progressing; now we can examine what is involved in climbs and what is involved in descents. The theories we've picked up in ground school are helpful here, since we can apply our knowledge of lift and drag to both our climbs and our descents.

Turns. Next we begin turns, medium ones at first. We'll put off steep turns until we are a bit more experienced.

Stalls. Now for the dreaded stalls. Almost everyone has been brought up to believe that stalling an airplane is a worse fate than selling your soul to the devil. This it most definitely is not. The stall is an airplane's built-in response to keep flying when you, the pilot, inadvertently try to stop it. It is dangerous only to the uninitiated; once you've learned about the stalled condition, it can be fun.

Spins. The FAA does not require student pilots to demonstrate their competence with spins, and many training aircraft are prohibited from intentionally spinning. This may be a mistake. A very large number of accidents are varieties of the stall/spin type; had pilots practical experience of how spins are caused and how they may be corrected, I believe there would be a lessening of this type of accident. At any rate, here we are going to deal with spins.

Circuits or pattern work. We now combine our knowledge of taking off and climbing turns, straight and level flight, and descending turns and landing into what we call circuits or pattern work.

Solo. Everything's been falling into the slot. One beautiful day your instructor says as you complete a series of perfect touch-downs: "Pull up here," and then he gets out. "Shall I get out?" you start to say, foolishly. "I want three circuits from you," he says, "and you can meet me back at the coffee shop."

HOW AIRPLANES WORK

Airplanes fly because of the properties of airfoils and the similarities of air and water. Consider for a moment the quality of air. Although we perceive the movement of air only in the form of wind, air has substance in the same way water does. If you wanted to extend the comparison a step further, you could consider the way fishes live in their watery environment and the way we live in our airy one. Oxygen is the common denominator, the fishes extracting it by means of their gills out of the water, while we, with our lungs, extract oxygen out of the air.

Air and water exhibit similar properties when they flow around surfaces of a particular shape. If you've ever held a spoon against water from a faucet, you'll notice that the spoon is drawn into the water. If you have a model airplane, try blowing smoke across the wing, and you will see for yourself the behavior of an airstream on the upper and lower surfaces of the model's airfoil.

Before we took off, we inspected the leading edge of the wing for any malformation. The reason for this is that when presented at a slight angle to the relative airflow, pressure will decrease on the upper surface of the airfoil. This is what we call lift, and it is controlled by two factors: the angle of attack and the speed of the airflow over the surfaces. Almost all parts of the airfoil contribute to the production of lift. (See Figure 2.1.) To deal more precisely with the effects of the forces an airplane is subject to, it is easier to simplify matters by saying that lift is generated at one point—the center of pressure. (See Figure 2.2.)

There is also a more ominous factor—drag. Like anything else that would move, an airfoil is subject to the forces of friction or inertia, which we term drag. This is a force of opposite polarity to the forward movement, and consequently drag works at right angles to lift. It may be alleviated, up to a point, by varying the angle of attack at which a wing is flown.

What happens is this: As we increase the angle of attack, we move the center of pressure forward; lift becomes more powerful, but unfortunately so does drag. After about 5¼ degrees of angle of

31

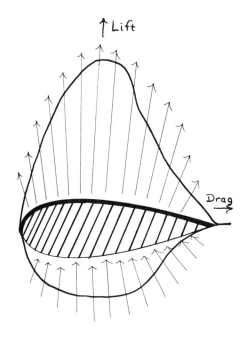

FIG. 2.1

Almost all parts of an airfoil surface contribute to its lift, but as may be seen from the sketch, the bulk of the lift occurs approximately one-third of the way back from the leading edge.

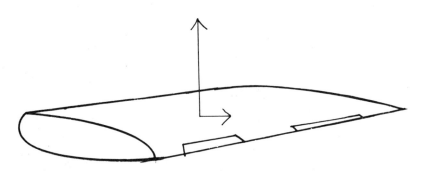

FIG. 2.2

It is convenient to show lift and drag occurring at a single point.

attack, the increase in drag becomes quite rapid until at about 15 degrees (up to 18 degrees on certain airfoils) insufficient lift is generated to compensate for it. (See Figure 2.3.)

Airfoil shapes are important in the generation of lift, as is actual wing area. The larger the area of the wing, the more weight it will tend to support at a given speed. The density of the air plays its part too; the more dense the air, the more lift a wing will generate at a particular speed.

FIG. 2.3

In normal flight drag is not too important a factor, but as the angle of attack increases, so too does drag, at a quite dramatic rate. At about 15 degrees, drag overcomes lift and the wing ceases to fly.

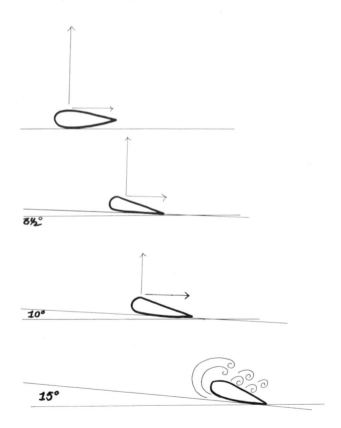

If you look at the diagram of airfoil shapes (Figure 2.4), you'll get a better idea of how this works. The very thin airfoil marked "supersonic" is designed to provide minimal resistance to the very rarified air at 50,000 feet and above, where supersonic aircraft fly.

These are the four forces an airplane is subject to: lift; its opposite, the weight of the airplane; drag; and thrust, which we balance against drag.

The Wright brothers mounted the engines of their flyer with propellers at the rear. As millionaire inventor Bill Lear once remarked, it was just as well that they did, or their airplane never would have flown. "Pusher" airplanes (See Figure 2.5) are actually more efficient than aircraft whose engines pull because the separate airflow created by the rotation of the propeller causes certain problems (to be dealt with in a moment). As far as engineering is concerned, placing the engine at the rear of the airplane is not really any more difficult than putting it at the front. But because we have become used to placing our engines at the front, design engineers don't really feel like bothering to make a change. (Except Jim Bede, whose extremely efficient BD-5 is a pusher, interestingly enough.)

It may be worthwhile to examine for a moment how propellers function. Without getting into heavy math, an aircraft propeller is basically an airfoil consisting of two or more blades. Like the wing, it is capable of producing both lift and drag, but to avoid confusion

FIG. 2.4

Airfoil shapes. From top to bottom: Fat airfoil provides good lift but excessive drag at high speeds; center airfoil combines good lift with reasonable cruise characteristics; bottom airfoil is for supersonic aircraft.

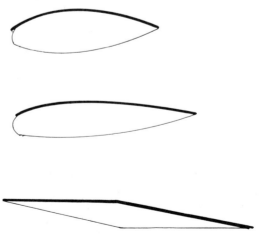

FIG. 2.5

The BD-5—a "pusher" type airplane

we talk of this lift and drag in terms of thrust (or torque) and torque reaction. If you think about how lift and drag is going to be distributed—bearing in mind that the outer edge of the prop moves much faster than the inner—you'll see that the angle of the blade decreases progressively from the center to the outside. It is not unusual for even quite small engines to have propeller-tip speeds approaching transonic speed, while at the center the speed may be only about 80 mph. (One of the ways for making propeller-driven aircraft quieter is to reduce the propeller-tip speed).

To understand how a propeller actually works, think in terms of a simple wood screw. The screw in one turn will penetrate just as far as the pitch of its thread will permit. A fine thread will give rather slight penetration—and this type of screw is more usually found for use with metals—while the coarse thread of a wood screw permits deeper penetration. A fixed-pitch prop is a compromise; indeed, on some small airplanes the manufacturers offer owners the option of having either a cruise or a climb propeller. The reason is that, if you can't alter the angle of the blades to the air, you may find at least a part of each blade in a stalled condition, with the unstalled part working rather poorly. As speed increases, the stalled part of the blade becomes unstalled, and soon the propeller can begin to work efficiently. At the top end of the scale, however, its efficiency may again begin to break down; beyond a certain speed the propeller, pitch being fixed, may be unable to do anything with the air immediately ahead of it because it cannot alter its angle of attack. And since it is an airfoil, it *must* work at an angle of attack to the air about it if it is to do its job.

By placing an engine at the front of the wing, we create a problem of balance. As stated in the first chapter, the fuselage is the lever to which we attach the tail to provide for the balancing of the engine and cockpit at the front.

We now have designed something that will fly, but in order to get it into the air we'll need to put some wheels under it. This is the undercarriage, which consists of two main wheels and a nosewheel, or in older aircraft two main wheels and a tailwheel. In the tricycle-geared airplane, the main wheels are placed behind the center of gravity, with the nosewheel forward. The airplane's wings are usually at a negative angle of attack so that they will not produce lift until the nosewheel is lifted from the ground. In tailwheel aircraft the main gear is located forward of the center of gravity; because the wings are already at a positive angle of attack, it would be possible to fly this airplane straight off the ground.

In order to operate our airplane we shall need some controls. We also could tidy up its shape by reducing drag caused by poor design, such as in the engine cowling, wheel gear struts, and so forth. Now we have an airplane.

As we are already familiar with the walkaround, let's take a look at emergency procedures. Although emergencies are extremely unlikely, provided we have made a thorough and detailed inspection before takeoff, they have been known to occur. Fuel failure—by syphoning, by water in the fuel, or by simply running out of it—is obviated by the proper walkaround: You visually check the gas for water contamination; you visually check the level of fuel in each tank; and you personally put the fuel tank caps back on. But suppose some part of the engine malfunctions, which does happen, even if very rarely.

Even though a part has failed, the modern aero engine will most often still provide sufficient power to enable you to make a safe landing. But in the event of a complete failure or fire, you must know what to do before it occurs, as time is a factor. A good pilot always has a plan for an emergency.

FIRE DRILL

We said that the treatment of engine failure and of fire was similar. The reason is that if we have an engine fire, we have to switch off the engine. Cabin fires, when they occur, are usually caused by electrical faults or careless smokers. Most modern airplanes have no restrictions on smoking in the cockpit, and ashtrays are usually provided. Still, ashtrays have been known to come unstuck and to spill their contents on the floor. Even fire-resistant carpeting can smolder, especially if there is a draft to provide additional oxygen.

A clue to electrical fires is usually the smell of smoldering insulation and the popping out of one of the circuit breakers. If the trouble is in the battery, the airplane may still be flown, though the master switch will have to be turned off. There are extinguishers now on the market that can handle both electrical and nonelectrical fires, and you may consider it prudent to invest in one.

The engine must, by regulation, be separated from the passenger compartment by a fireproof bulkhead. Still, serious fires can develop when either oil or fuel lines break, and the slipstream effect can actually bring flames aft. If an engine fire should develop while in the air, you need to recognize immediately whether it is a fuel or oil fire. Burning oil produces a heavy dark smoke, whereas gasoline will produce very hot flames.

With the engine and the fuel tanks switched off, the flow of fuel to the fire is reduced to what is left in the fuel lines. Sometimes it is possible to blow the fire out by diving. But oil fires present a more difficult problem; even though you may have shut down the engine, the propeller, as it is turned by the wind, will continue to turn the engine over. The technique here is to slow the airplane down (just short of the stall) to reduce the flow of air through the engine cowling. Smoke—and flame—can usually be kept away from the cockpit area by slipping the airplane, and occasionally this will also help to blow out the fire.

Finally, there are ground fires. Usually these fires start around the carburetor inlet, and the quickest way of dealing with them is to starve them of air. You must still deprive them of fuel, which means you must close the throttle, switch off fuel to the lines, and lean out or reduce air/fuel mixture to the engine. Unfasten your seat belt and keep cranking. The cranking is to suck the fire into the engine. (Of course, if the fire's too big to suck into that little engine, perhaps you should leave the airplane.) Don't be too intent on using the extinguisher on the engine, because if the engine sucks it all in you'll need to take apart and clean the engine before you can use it again.

The formula for fires, therefore, is throttle CLOSED, fuel OFF; let the engine stop running under its own power, and then turn the ignition off.

In the event of an engine failure due to faulty fuel flow the engine will have stopped before you begin your emergency landing. It is important to make sure that everything is turned off once you have decided there is no way to get the engine running again. Here, however, you should check the carburetor heat—perhaps icing is the problem. (2) Is the air/fuel mixture too lean? (3) Is fuel flowing from the tanks? You can sometimes overcome serious carburetor icing by revving the engine up and down and by switching the ignition off and on. This will cause backfiring, which can break up ice (and even melt it) quite effectively. It is not very good for the engine.

The key to any air or ground emergency is *not to panic*. You will be surprised how efficiently and speedily you can think if you avoid panic and have a master plan prepared for whatever eventuality, which is, after all, the purpose of all emergency drills.

AT THE CONTROLS

We've already studied some of the theory behind the controls. We know that the ailerons work because they interfere with the flow of air across the airfoil surface. We know that we balance our turns with the rudder, to counteract the yaw produced as we use the ailerons. And we know—at least in theory—that the elevator works for us by controlling the angle of attack. Soon we will go into the air and see just how these work in practice.

But before leaving the ground, let's have a look at the diagrams to see what occurs when we move the controls. (See Figure 2.6.) First, the ailerons. For a left bank, we turn the wheel to the left, and the left aileron goes up as the right one goes down. There's just a touch of yaw when we do this in the air, so we apply just a touch of rudder in the direction of our turn to balance it out. For a right bank we turn the wheel to the right, and this time we'll add a touch of right rudder to coordinate the turn properly.

Now what about the angle of attack? With the wheel back, we increase the angle of attack. We also increase drag and cut back on our speed. With the stick forward, we lessen the angle of attack, lessen drag, and thus add speed.

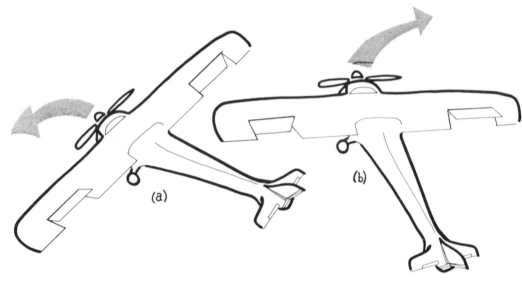

FIG. 2.6

(a) Up left aileron, down right aileron produces a bank to the left. (b) Up right aileron, down left aileron produces a bank in the same way but in the opposite direcon of (a).

Now let's get airborne. First, the ground checks (see page 9) and then the cockpit checks. You can make the takeoff this time all on your own. Cleared for takeoff, ease the throttle forward, keeping the airplane along the center line of the runway with light pressure on the rudder pedals. Glance at the airspeed indicator, and as it comes through 60 mph, ease in some back pressure on the control wheel. Nosewheel up; as the speed continues to increase, the aircraft lifts itself off the ground. Keep the speed at an indicated 80 mph and we'll be able to watch for traffic. Now you can start a gentle turnout, as we are leaving for the practice area; I've already requested a nonstandard departure from the field.

TURNING

We are now at cruise altitude, so lower the nose to cruise position—keep it there—and when you reach cruise airspeed, ease back the throttle to 2400 rpm and trim out the weight on the controls with the trim wheel. Try a gentle turn to the left. Remember, just a touch of rudder as you move the control to the left. See the ball in the turn and bank? Dead center. See if you can do that to the right now.

In order to turn to the left, we must roll or bank the airplane to the left. The aileron is our roll control, so we apply a gentle pres-

The aileron

sure to the left. As long as we hold the pressure, the airplane will continue to roll. When we release the pressure the airplane will maintain the angle of bank.

Try it once more: Bank, and it wants to keep on banking, so you automatically neutralize the controls and it stays there. The reason for this is that as you bank, two new components are introduced into the situation. First, there's the tendency for the airplane to continue to move sideways, which will, if not checked, appear to lead the nose of the airplane in the direction of the bank. This can develop into a spiral. Second, one wing is now moving through the air faster than the other.

Now let's develop this idea a stage further. We've noticed that if we use a judicious amount of rudder at the beginning of our turn, it is balanced. Let us now do it the wrong way, just to see what happens. We'll start with the rudder. Step fairly briskly on the right rudder pedal—don't touch the controls. The nose starts to swing around to the right, the left wing now banks because it is moving faster and is producing more lift, even though you are not touching the ailerons. If you kept it like that you would end up in a spiral. The same thing happens on a yaw to the left—the right wing, moving faster, develops more lift and tries to bank.

Now centralize the rudder and move only the control wheel for an ordinary left turn. Look at the ball in the turn and bank; you will discover that you are actually slipping slightly to the left. If you leave it like this, you will again start a spiral dive. Try it to the right, still with the rudder centralized. You get the same effect, but in the opposite direction.

40

From this it can be seen that the rudder balances out the yaw effect when using the ailerons and the ailerons balance out the roll effect of the rudder. When making small corrections in course when flying instruments, or simply picking up a slightly dropping wing at slow speeds, the rudder is the most useful and efficient control to use. But the basic controls for banks, turns, and course changes are the ailerons.

You may have noticed that when you banked the aircraft, the nose dropped. By diverting lift from the vertical you've upset the balance of forces that keep the plane in the air. The nose wants to fall to restore them. If this tendency is left unchecked, a spiral dive might result. We overcome this tendency by the simple expedient of back pressure—enough to keep the nose where it belongs.

The fact that we are holding back pressure tells us that we are at a higher angle of attack. Or, stated another way, we are closer to the stall. More about that later. If you look at the diagram (Figure 2.7) you'll see why.

Now how about trim? Airplanes can be flown just by using the trim wheel for the control of the angle of attack. The fore-and-aft trim removes the control pressures from the control column. If

FIG. 2.7

When an airplane is turning, its outer wing has to travel further than the inner wing. The additional lift created by the extra distance the outer wing must travel in the same time creates an instability.

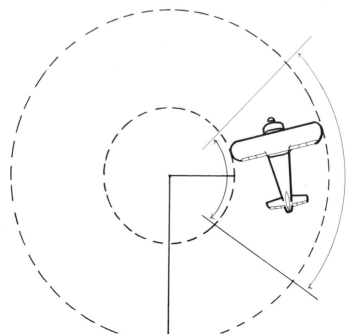

Conventional tail assembly, consisting of fixed stabilator at the front (in the horizontal plane) with a moveable elevator at rear. The trim device is located on the trailing edge of the elevator.

Shown here is the fully flying horizontal tailplane of a Beechcraft. The entire horizontal surface moves up and down, to change the angle-of-attack of the aircraft. The thin section on the after end of the tail provides trim forces, and moves more acutely in the *same* direction as the main plane.

you're climbing or descending, you "dial in" the correct amount of trim needed to hold the airplane at its appropriate attitude. In most modern airplanes the trim wheel interconnects with a small movable surface on the elevator, which looks as if it might be a mini-aileron but is actually more like a mini-flap that can move both up and down. Having adjusted the elevator to the desired angle of attack, you now adjust the trim tab to take out the pressure of holding the elevator with the control column for the particular angle of attack you have selected. It provides a very small out-of-balance force to the elevator itself. Curiously, this has the effect of making the elevator hold itself in the position to which it was set. If you have a look at the illustration (Figure 2.8), perhaps you will see that the idea is not as crazy as it sounds. In larger aircraft you will come across a rudder trimming device, which is very pleasant for taking out rudder pressure—except during takeoff when normal position is neutral. (But not in a P-51.)

American aero engines turn their propellers from left to right, as seen from the cockpit. The British arrange their propellers to turn from right to left (and they still drive their cars on the left-hand

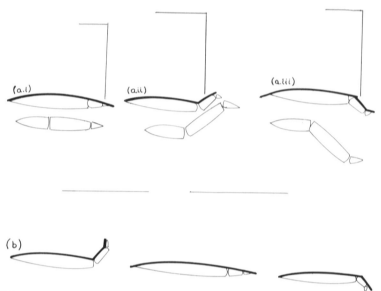

FIG. 2.8

Trim tabs are basically devices that keep a control surface in the position to which it has been adjusted. In (a.i) the trim tab is the little device at the end of the control surface in level flight. In (a.ii) the elevator has been raised for a slight increase in angle of attack, perhaps to provide for a climb. In (a.iii) the position is reversed, perhaps for a descent. In (b) the situation is shown with a full flying tail surface.

side of the road). Airplanes have either a left- or right-hand turning tendency as a result of the revolving of the propeller—torque—and the airplane designer must counteract it. There are a number of means available to counteract torque; designers usually employ one or more, including the following:

- Setting the vertical fin off-center to provide a permanent force in opposite direction to torque.
- Offsetting the engine installation in relation to the fuselage so that the propeller pulls more to one side and counteracts torque.
- Fitting a ground-adjustable tab to the trailing edge of the rudder.
- Fitting an adjustable trim tab to the rudder. This may be adjusted in flight.
- Fitting a spring arrangement (occasionally adjustable in flight) to pull the rudder control bar in the appropriate direction.

This means that you will have to counteract torque when it appears. It can appear also as P-factor. P-factor is the effect of torque on the fuselage of the airplane, causing the airplane to yaw in a direction opposite to the turning of the propeller. It is most noticeable at high angles of attack with high power settings and requires

43

—in American airplanes—a fairly heavy degree of right rudder to counteract. It occurs when the flight path of the airplane is not exactly at right angles to the rotation of the propeller. Engineers describe this as asymmetrical lift developed by the propeller airfoil, in which the downgoing blade is producing more lift (thrust) than the rising blade. Your instructor will demonstrate this condition for you.

STRAIGHT AND LEVEL FLIGHT

Straight and level flight is next on our agenda; although it sounds perfectly simple, it takes a little getting used to. The reason is that we have to adjust our eyes to the feel of straight and level flight in relation to the things outside. One of the simplest methods of judging this is to note where the top of the instrument panel meets the horizon, and how the wing tips appear to relate to the horizon in level flight. Or, as you may hear your instructor put it: "You must get used to the attitude of the nose relative to the horizon."

The trick to straight and level flight is to let the airplane settle down on its own. Obviously if the airplane continues to climb once you've trimmed it, you'll need a little more trim, and if it continues to descend, you'll have to add some more of the opposite trim.

Once an airplane is in reasonable balance, you can check phugoid oscillation very quickly. Phugoid oscillation is a long-term characteristic of imbalance in longitudinal stability. Put in simpler terms, it's the up-and-down movement of the center of mass in relation to the motion of its longitudinal extremities. Put more precisely, it's the slow up-and-down motion of the nose. What happens is that the aircraft hunts for a proper balance for angle of attack and speed. The mistake most beginners make is to alter the airspeed to compensate. This is a mistake because we know that the angle of attack can be made to work for us; at a given power setting, properly trimmed, we should automatically have a given airspeed. So it is a question of relating the airspeed to the power setting, or again—"attitude of the nose relative to the horizon."

CLIMBS AND DESCENTS

Smoothness applies to piloting an airplane in much the same way as it applies to chauffering fine cars or riding horses. Just as a good driver smoothly accelerates his car to cruising speed and brings it evenly and gently to a halt so that the passengers aren't jerked around, the good pilot smooths his transitions across the skies.

A good number for both climb and descent is 500 fpm on the VSI. As mentioned earlier, at this rate, pressure on the eardrums is minimal, and airspeed and heading may be easily maintained. If you have looked in the owner's manual for the aircraft in which you are learning to fly, you may have discovered that a special speed is mentioned, usually listed under the maximum *rate* of climb data. A typical trainer lists this at 670 fpm at 76 mph indicated airspeed. However, this speed and rate of climb is not the best *angle* of climb. Your instructor will explain it in due course, and it is noted later here.

The important thing to be aware of when climbing or descending is engine temperature. Climbing at too low an airspeed tends to overheat the engine if one isn't careful, and you won't change altitude very quickly if your airspeed is too high. Obviously the thing to do is to take advantage of the best angle of attack, which will optimize the lift/drag ratio. This will be somewhat faster than the maximum rate-of-climb speed mentioned and will produce about a 500-fpm climb in light trainers.

From cruise speed you will automatically add full throttle to start your climb, easing the stick back until you have achieved the airspeed you want—let's say 80 mph. Keep the wings level and use a little rudder to counteract torque. If you will be climbing for a long time, check in front of you very carefully by turns to the left and right. Upon nearing the altitude you want, a good practice is to start easing forward just after you pass through the desired altitude so that you have a little extra in hand, maybe fifty to a hundred feet. Now as the airspeed increases to cruising speed, reduce the throttle setting to your former cruising rpm. Trim for level flight and with slight forward pressure get exactly on to your altitude. Retrim as needed.

Descending is the same procedure in reverse, with the exception that you may reduce power to almost zero if you want to. Whether you make a powered descent or a gliding descent, you should remember to use carburetor heat—carburetor icing can occur in certain weather conditions, even on bright sunny days.

Turning once more to the owner's manual, you will discover a chart that shows Maximum Glide, which is the speed at which the airplane will glide most efficiently with the propeller turning. For our particular trainer, this speed is listed as 70 mph; from 6,000 feet we would be able to glide 10 miles. That represents a glide ratio of slightly better than 10:1. Sailplanes these days have glide ratios of better than 50:1, which means that from the same altitude they would be able to glide five times as far.

But what can we do if we don't want to glide so far with our powered airplane? Control of the glide path is a useful skill, especially when practicing emergency or precautionary landings. (See Figure 2.9.) It is not usually taught until the student has demonstrated that he is proficient in maintaining the proper airspeed for the most efficient glide. The essence of glide control seems contradictory at first sight: if you want to go down more steeply, raise the nose.

In the same way that you are able to achieve your maximum rate of climb at a slower speed than your normal climb speed, so you can play with your glide speed. There is not much advantage in gliding at a higher speed than the one recommended because as you get nearer the ground you'll have to lose the speed. And the disadvantage of gliding at a slower speed than the most efficient gliding speed is that your rate of descent will have to be slowed as you near the ground. Your instructor will show you how this works; by knowing how to do this, you will be able to make corrections.

FIG. 2.9

It's impossible to stretch your glide, since there is an optimum distance an airplane will fly from a particular altitude. Plane a is gliding too slowly (60 mph), while plane b is gliding too fast (90 mph). Plane c is just right (78 mph)—he's going the furthest.

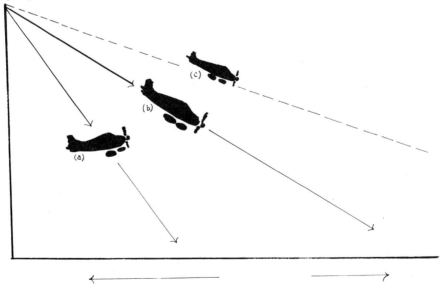

Powered descent is the more usual means of changing altitude in flight. What happens is that you bring the throttle back from cruising power until you establish a 500-fpm rate of descent. In our trainer this happens to be 1750 rpm on the tachometer. You trim the aircraft to maintain the rate of descent. As you approach your new altitude, increase power up to cruising speed; retrim for straight and level flight.

For landing we usually use a mixture of the pure glide and the powered descent; by increasing or decreasing power we can control our approach very precisely. We achieve this by first setting up a good approach speed. To increase lift for our slow speed capability, we start by slowing down the airplane and then add some flap; a slight amount more of flap will enhance the lift of the wing at slow speeds. Most probably, your first landings will be made without the use of flap because when flap is added, the attitude of the airplane will usually change. This can cause confusion for beginners, who have enough problems getting everything into sequence. Anyway, establish a reasonable airspeed for your descent, between 70 and 80 mph with the flaps up or between 60 and 70 with the flaps extended. You should have already added carburetor heat and loosened the throttle friction nut so that you can apply more power as needed to control your descent. A good rate of descent on final approach is a sink rate of about 300 fpm, but let your instructor explain this to you.

Another method of losing height, which we touched on earlier, is the slip. Until flaps became a regular feature on most aircraft, the slip was employed as a means of losing height rapidly, even during the final approach. The advantage of the slip is that there is no increase in airspeed, which would normally need to be decreased were one to dive to reduce one's altitude.

When the wing is lowered, you will remember, yaw will develop toward the down wing. In a slip we deliberately lower the wing and then prevent the resulting yaw by delicately applying rudder in the direction opposite to the down wing. The effect of this cross-controlling is to alter the path of descent in the direction of the lowered wing, while slightly adding to the actual airspeed. When we use a slip to lose height for a landing, we actually yaw the nose around before starting the slip so that during our slip we are still lined up with the runway. Actual sideways speed may be as much as 25 mph, so you see it is very important to line up the flight path before starting.

The slip is an excellent method of losing height without too much gain in airspeed, but because cross-controlling can lead to

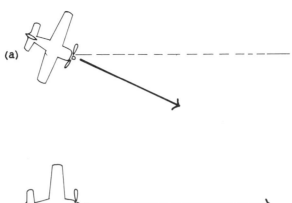

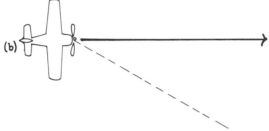

FIG. 2.10

The sideslip (a) and the forward slip (b). The heavy arrows indicate the ground track.

serious problems if taken too far, you should not experiment with slips without your instructor. The forward slip and the sideslip are performed in exactly the same manner, the only difference being the actual track of the aircraft over the ground. Figure 2.10 explains this.

CLIMBING AND DESCENDING TURNS

Here's a slightly more complicated maneuver than a level turn—the turn while climbing or descending. In a level turn the only problem you need to worry about is that the outer or upper wing in a bank (because it is moving faster than the inner or lower wing) develops more lift. (Remember, the outer wing has to travel a greater distance in the same time.) But because shallow turns are not very efficient—we use them only for minor course corrections—we must investigate what might be termed the medium turn.

Look at the turn coordinator for a moment: There's a mark to the left and right of the mini-airplane's rudder. If the panel in your aircraft has the old-fashioned turn-and-bank, there is a mark to the left and right of the needle. This mark indicates what is called a "standard rate turn"—a turn in which the airplane reverses its course through 180 degrees in a period of 60 seconds. That's a rate of three degrees per second. An interesting point about the rate of turn is that it is dependent upon the airspeed and the angle of bank. The faster the airplane is traveling, the greater the angle of bank must be for a given rate of turn. Low speed, therefore, requires a small angle of bank for a standard rate turn, whereas a higher speed requires a greater angle of bank. Your training airplane, for example, traveling at 100 mph would require a 15-degree angle of bank. A business jet aircraft (bizjet) traveling at 500 mph would require a 55-degree angle of bank for the same rate of turn—quite a difference.

Another factor must now enter our calculations—the redistribution of lift. In level flight the wing area develops sufficient lift to compensate for the weight of the aircraft. Let's take an easy number and say that the aircraft weighs 2,000 pounds. As we bank, we start adding new forces to prevent the airplane from skidding or losing attitude in the turn. First, we have to support the weight of the aircraft, and then we must direct the craft in the direction of the turn. This second force is called centripetal force, and the tighter your turn for your airspeed, the greater this force must be. Obviously, something is lost, and this is airspeed. In a standard rate turn you will lose about 5 mph.

If you take a look at the diagram you will see that when we enter a 30-degree banked turn, we have to add a turning force of slightly more than half the airplane's weight—1,100 pounds—and must increase the lift to 2,250 pounds. If we steepen the bank to twice this, we would require 4,000 pounds of lift, plus a turning force of 3,500 pounds. To maintain altitude, we get our additional lift by increasing the angle of attack with more back pressure on the control column, and the result once more is decreased airspeed, because of the greatly increased amount of drag.

In order to prevent the angle of bank from steepening, we must neutralize the turn once it is established by centering the controls canceling out the aileron force.

Because we must redistribute our lifting forces in turning, we apply an extra load on the wing. We compensate for this by adding additional power as we go into steep turns. At the same time we must be wary because the greater the angle of bank, the higher our stalling speed. If you will remember, we said that we had to ease back the control column in order to increase the angle of attack. But, as the angle of attack is increased, the lift/drag ratio also increases, in favor of drag. Unless we have plenty of power available to counteract the drag, we shall stall. And in very steep turns this is what can happen—most light aircraft simply don't have enough power available to compensate for the extra drag.

49

For this reason the wise pilot is fairly cautious about banking his aircraft at low speeds while attempting to maintain altitude. In this condition, without adding plenty of power, he could find himself in a stall/spin type of predicament. If he uses common sense, all will be well; once again, this is something to explore with your instructor. When you get the hang of turns, especially steep turns, you will revel in your new-found abilities. A good rule to follow if a stall seems likely is to immediately ease off back pressure—that is, in effect lowering the nose.

Now let us apply this to climbs and descents. All we do is to increase speed in the climb and increase power during descent. If you think about it for a moment, you'll see why.

During a climbing turn the aircraft tends to increase its angle of bank much more than in an ordinary medium turn. The reason is easy to see, as the airplane is virtually moving up a spiral. The inner or lower wing is, therefore, flying at a lower angle of attack in relation to the outer or upper wing. We have discovered that the best rate of climb was obtainable at a certain angle of attack with a certain amount of power. Since we are upsetting the balance of lift between the right and left wings by turning while we are climbing, we'll need more power to obtain the angle of attack we want, as we already know that we must lose some airspeed in order to add to our lift.

We should add speed in our climb—say 10 mph. Perhaps we should say that we should limit our angle of bank. Indeed, in a climbing turn the angle of bank should be limited to not more than a standard-rate turn, as steepening the bank requires even more power. So in climbing turns the rule is to add more power than is needed for a level turn and to limit the angle of bank to not more than is required for a standard-rate turn. Keep the ball of the turn and bank indicator centered all the way up.

During the descending turn you will discover that the bank must be held throughout the turn. The reason is that the inner wing, while traveling a shorter distance than the outer wing, is actually flying at a greater angle of attack, which more than compensates for the outer wing's travels. Once again the angle of bank is kept relatively shallow.

A variation on the descending turn is the slipping turn, used to reduce altitude during a descending turn. It is occasionally useful for landings at fields where a straight-in approach is obstructed by telephone or power company wires.

STALLS

We saw earlier that beyond a certain point, when the angle of attack becomes excessive, drag increases faster than lift to an extent at which the air no longer flows smoothly over the wings. The process is a gradual one, with the stall beginning near the wing root and slowly extending out to the wing tip. (See Figure 2.11.)

Modern technology provides two methods for delaying the stall. The first is the application of a leading edge cuff or slats, which are shaped to encourage the air to flow smoothly over the wing into very high angles of attack such as on the English Tiger Moth. The second is the stall fence, a thin, fore-and-aft section of metal on top of the wing near the ailerons that inhibits the tendency of stalled inboard air moving out to the wing tips. Neither measure prevents the stall. The first delays it and enhances control within the stall; the second ensures good aileron response in the stall.

Since a stall is caused by using too high an angle of attack, the solution must obviously be to reduce the excessive angle of attack to one that is reasonable. And since a too high angle of attack is involved in causing the stall, it must be understood that the condition of stalling has nothing to do with airspeed. It's important to

FIG. 2.11

(a) At level flight, slow speed, aircraft is in incipient stalled condition. (b) Moment of stall is occurring, and the nose of the aircraft is about to drop.

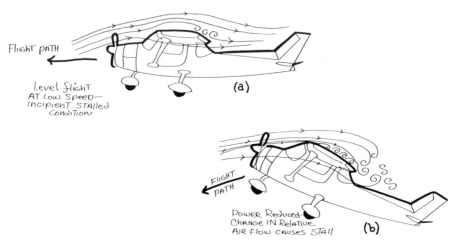

Flight path

Level Flight
At Low Speed—
Incipient Stalled
Condition

(a)

Flight
Path

Power Reduced—
Change In Relative
Air Flow causes Stall

(b)

understand this point, as it is perfectly possible to stall an airplane by suddenly pulling back on the controls when you're cruising—if you don't pull the wings off first.

However, before a stall occurs, you usually get a warning in the form of airframe buffeting—a shuddering of the aircraft—and the reedy shriek of the stall-warning device. Still, not all aircraft are thoughtful enough to warn you of an impending stall. A good practice to follow the first time you fly a different make of airplane is to find out what its stalling characteristics are like.

We're at 4,500 feet now, and we make a couple of clearing turns through 180° in each direction to make sure that we're not in anybody's way. We've already checked our harnesses and the doors, that there are no loose articles in the cockpit: also, that we know our position relative to an obvious landmark and that we are not over a built-up area.*

Now I'm going to show you how a power-off stall works. First we apply carburetor heat; throttle is back, and I'm keeping the airplane straight but raising the nose above the horizon to reduce the speed. If you watch the airspeed indicator you'll see that it is falling. If you handle the controls, you will notice that they are becoming less effective. I keep adding a little more back pressure to the control column, and now you can feel the beginning of the prestall buffet, as if we were riding in rough air. Now the stall indicator reed begins to quaver—what a noise! The airplane is now very close to the stall. Notice our rate of descent even though the nose is pointing up.

On this aircraft the stall break, with the nose dropping forward, is very gentle. To recover from the stalled condition once the nose has dropped, we move the control column forward and increase power as the nose comes up to the horizon, easing out of the gentle dive.

Our next task is to investigate the power-on stall. Because our engine is providing us with plenty of thrust, the onset of the stall is delayed; consequently we can get the nose much higher before the break.

Because there is a greater tendency for the stall effect to be much sharper in a power-on stall, we must be even more careful that we

---

* HASEL checks, a useful mnemonic: *height*—sufficient for the exercise; *airframe*—brakes off, flaps as required, gyros caged; *security*—harness and hatches secure, no loose articles; *engine*—fuel pump (if any) on, carburetor heat as needed; *location*—out of controlled airspace (not over a town or other built up area, not over an airport or over other airplanes); make an inspection turn to "clear" the area first.

use opposite rudder rather than the aileron in picking up a dropping wing. The reason is that if we use the aileron to lift up the wing in the semi-stalled condition, we will make matters worse; the wing becomes fully stalled and thus all set for the incipient spin, which, if not corrected very quickly, develops into a spin proper.

The reason this happens is—as you may remember—that the ailerons move in opposite directions; that is, if we move the wheel/stick to the right, then the right aileron goes up and the left one down. Now the down-going aileron causes far more drag than the up going one—a condition known as aileron drag—and since in this case it is on the wing that has dropped, i.e. which is partially stalled already, this extra drag is enough to cause a near complete stall on the down-going wing. Thus you are all set to spin—so, *always* use rudder to raise that dropped wing.

If the left wing has dropped then right rudder causes yaw, which in turn causes the airflow to be greater over the stalled left wing than the right one, and the wings therefore become unstalled.

In the power-on stall we reduce power a little once we're through the stall break, as we don't want to increase our speed unnecessarily in the resulting dive.

The other situation in which stalls occur is during a steep turn. Your instructor will demonstrate this by going into a steep turn without increasing the power. The radius of the turn is now shortened by easing back on the controls. If you watch the airspeed indicator you'll see that as the stall buffet starts, the airspeed is much higher than it was in both the power-off and power-on stalls. Recovery is easy—just ease off the back pressure and level the wings.

SPINS

In the early days of flight there were many mysterious accidents. Airplanes unaccountably tumbled over in flight and plunged to the ground, falling through the air that until moments before had supported them. Gradually it was found that there was a certain predictability about this type of accident, and several brave pilots attempted to discover what it was that produced this effect and how it might be avoided.

These accidents were spins. For some reason it was impossible to come out of a spin, and the only way of saving oneself was to use a parachute and jump out of the stricken airplane. (Paradoxically, British pilots during World War I were not permitted the use of parachutes. The authorities declared that it would tend to make them cowardly and that they would desert their machines.) Finally, some unsung hero discovered that you could stop the rotation of the spin by using the rudder in the direction opposite to it and that you could break the stall part of the spin by easing the stick forward. Provided you had not lost too much altitude to come out of the dive, you could save yourself.

Today, spins are a regular part of demonstrations and aerobatics, and they really are fun to do. Accidental spins occur during several flight conditions and are always the result of mismanagement of the controls or lack of knowledge. Spins can result from making a gliding turn at too low an airspeed, and because loss of altitude is considerable, it is this sort of spin that claims victims—especially on the turn from base leg to final approach.

We want plenty of altitude, as altitude loss in a spin is great. Again we make sure our harnesses are on firmly, that the aircraft doors (and the baggage door, if there is one) are shut and locked, and that there are no loose articles around. Now we proceed as if we were going to do a power-off stall. That is, we apply carburetor heat, reduce throttle, bring back the airspeed, and raise the nose. As we approach the stall, we apply sharp right rudder. The right wing drops violently, the airplane slips down toward it, and we are in a condition known as autorotation. (Spiral dives are different because the controls remain normal and the airspeed increases; in a spin airspeed does not increase and the aircraft remains in the stalled condition.

Figure 2.12 shows what is happening. If you examine it you will see that when we yawed the airplane with that brisk application of rudder we unbalanced the airflow over the wings. On the upward wing the angle of attack was reduced, whereas the angle of attack on the downward wing was increased, thus fully stalling it. But the upper wing, because of the lesser angle of attack, is now only partially stalled, or may not be stalled at all. One wing fully stalled and one wing with a little bit of stall produces a condition in which the increased drag of the stalled wing and the weathercocking action of the nose—and thus the remainder of the airplane— maintains the effect of yaw. In effect, the airplane itself becomes a miniature gyroscope.

A number of variables affect the spinning characteristics of an aircraft. Airplanes with short fuselages and long wings tend to spin rather fast and can be difficult to recover. Conversely, aircraft with long fuselages and short wings tend to spin slowly and are easy to control. Load also plays a part. If an airplane is tail-heavy, you can

54

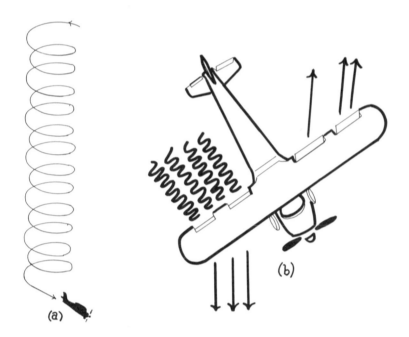

FIG. 2.12

(a) This shows the path described by a spinning airplane. (b) This shows how the spin occurs: One wing is fully stalled while the other is providing lift.

have considerable problems in a spin, as the spin will tend to flatten or raise the nose from the nose-down position, and you may even have to use differential throttle and jerk the controls back and forth before you can get out of it. (If you can get out of it.) It is therefore very important before attempting the spin exercise to pay careful attention to the weight and balance of the aircraft before you start since some aircraft have a restricted fuel load if aerobatic maneuvers are to be carried out. Your instructor will explain this more fully, and under no conditions should you attempt to practice spins without prior instruction.

Now to stop the spin. First, as we used right rudder to start the spin we'll use left rudder to break the rotation. As soon as rotation stops centralize the rudder. Now the aircraft is back in an ordinary stall; a slight pause and then ease the control column forward, and we are now in a dive with the airspeed building rapidly. Now, very, very gently—we don't want to produce a secondary stall—we ease back on the control column and bring the nose up to the horizon, once the nose is above the horizon, apply power to regain our original altitude, noting the height loss. If you look at the altimeter you'll notice that we lost nearly 2,000 feet, which is a very neat way to lose altitude in a hurry, though not very kind to yourself or your passengers.

The mechanics of the spin and the stall are easily learned. However, it should be learned very thoroughly, for harm can come to a person who is unaware of the dangers. Don't think that spins will come jumping out of the air at you on rainy days—you have to promote them. And as Wolfgang Langewiesche so neatly puts it: "A spin is nothing but a fancy stall; one side of the airplane is stalled, the other is not, and therefore the airplane sinks down twisting."

Spins are not required by the FAA for the private pilot certificate, but in your own interest ask your instructor to demonstrate the following: (a) an ordinary spin from straight flight and its recovery; (b) the incipient spin and its recovery; and (c) the spin from a turn. This last is the one that can get you into trouble, and it is worthwhile learning about it, how to recognize the condition in which it can occur, and what you should do to avoid it—unless, of course, you're simply practicing. One last word on the subject— don't practice stalls or spins unless you have plenty of altitude in which to recover. Remember, one of the most useless things to a pilot is the sky above him.

3

From Solo to Flight Test

There are two reasons why emphasis is placed by flight schools on pattern work (circuits, touch-and-goes). The first is that in terms of controlling an aircraft, a single circuit includes all that is required for the most elementary flight as well as for the most complicated cross-country flight—the takeoff, the climb to cruise altitude, the descent, and the landing. The second reason is that it gives you, the pilot, an opportunity to practice landing without wasting your time.

Someone once calculated that the average student pilot spends no more than thirty minutes of his total flight training in actually landing an airplane. Look at it this way: Takeoff takes about 20 to 30 seconds before you are airborne. You climb to the required altitude, and once in the pattern, check your height, speed, and tachometer. You then trim at the start of the downwind leg, make your prelanding checks before turning onto base leg and prepare the airplane—while you're still on base—for your landing approach. The approach may take a couple of minutes or more depending on how far out you went on your downwind leg and how much traffic there is. Now for the landing: Calculate that it starts when you break your approach glide; you'll find that from that moment until the wheels touch down, you will have spent at most some 20 seconds. When you consider that you have to get the airplane back up into the pattern, it's not surprising that landing takes no more than about one-twentieth or less of your pattern time. (See Figure 3.2.)

57

FIG. 3.1

Piper's Flite Liner is the low-wing alternative to Cessna's high-wing 150.

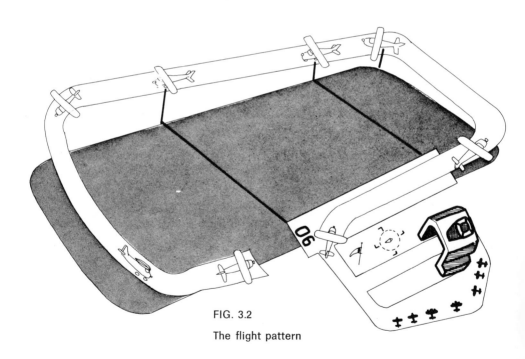

FIG. 3.2

The flight pattern

There are two or three easy ways to get landings right, and most pilots have their own favorite methods. Perhaps the most important point of all is to learn to use perspective as an aid. If you have the airplane in the proper position and glide angle relative to the runway, you will notice the ground gently moving both toward you and up to you. This is the key. Any change in your control input will alter the relationship between the two rates. If you ease back on the stick, the upward motion of the ground will decrease and may even be reversed. This perspective helps you determine what's really happening to your airplane.

Make a brief check of your instruments at intervals, as you should be looking outside most of the time. Your airspeed and rate of descent are the two items you should monitor most closely. Let's assume an airspeed of 70 mph on our approach and a sink rate of around 350 fpm. It's important to get these nailed down right at the beginning of the approach. A good landing is like a beautiful symphony—it must start right, right from its beginning.

With your approach speed at 70 mph and a comfortable 350-fpm rate of sink, visually gauge the two rates at which the ground is approaching you. We are not using flaps, but many instructors feel that students should make use of them from the start (the FAA seems to prefer landings with flaps too). It seems to me that for many students the use of flaps on the first several landings— because of the degree of pitch change to the aircraft—is somewhat confusing. So for our purposes we'll leave flaps out of it until we have landing technique correct. (See Figure 3.3.)

FIG. 3.3

This illustrates the effect of flaps on descent and range. Airplane A is using no flaps; airplane B has 20 degree (half) flaps; and airplane C has full flaps.

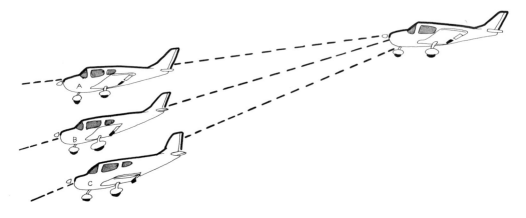

FIG. 3.4

(a) A normal landing; (b) the Crunch; (c) the Balloon.

Because we are in a training aircraft, we'll check our glide ("flare") at some point below thirty feet above the runway. (In a bizjet we'd check it a little higher.) Gently easing back on the control column stops the sink and cuts the speed, reducing the motion you see. Now the sink begins again and, as it starts, you very gently ease back just a little more. Glance at your airspeed; you don't want to stall. This cuts the sink, but it no sooner stops than it starts again. Very gently, give a little more back pressure on the control column, keeping the airplane off the ground. The sink stops, then starts again, and once again you apply a touch more back pressure trying to keep the aircraft just off the ground for as long as possible. And then you touch down perfectly. (See Figure 3.4.) Ease the nosewheel onto the runway and apply the brakes gently. Your touchdown speed was about 60 mph.

The emphasis is on *gentleness*. The pilot, like a skilled horseman, must have gentle but firm hands. Gentle, because an airplane, like a horse, is actually very responsive; firm, because, like a horse, an airplane needs to know who's in charge.

Now let us break down the pattern into its integral parts.

60

TAKEOFF

In order to reduce wear on the tires and to get off the ground in the shortest space possible, always take off into the wind. (There are exceptions, such as runways where you must land in one direction and take off in the other because of an obstruction, but these do not concern us here.) The reason for taking off into the wind is that wind contributes to getting the airplane into the air. This is how it works.

We established earlier that in order to provide lift the wings of the airplane must be moving through the air at a certain minimum speed. Below this the wings cannot do their job. According to the Owner's Manual for our trainer, we raise the nosewheel at 55 mph and use a climb speed of 70 mph. These speeds are relative to the wind. In other words, if we're taking off into a 35-mph headwind, our actual groundspeed when we raise the nosewheel would only be a little over 20 mph, while our apparent climb speed would be 35 mph. (If we were actually flying in these conditions, we'd increase our actual speed as a safety factor.) Now suppose we decide not to take off into the wind but with the wind, assuming that the wind (now a tailwind) has dropped to 20 mph. You will see that we cannot raise the nosewheel until we are actually going 75 mph (55 mph + 20 mph), and our speed over the ground is going to have to be better than 90 mph for us to climb. So it simply isn't worth doing this the wrong way—it's counterproductive. (This also applies to landing downwind for the same reasons—though a downwind landing can be even more uncomfortable, if not downright dangerous.)

Raising the nosewheel brings the wings to a reasonable angle of attack relative to the wind, and the airplane will very quickly lift off the ground. Before we begin our climb, lower the nose momentarily by easing the stick slightly forward to permit the airplane to build up climb speed in ground effect. From the beginning of the climb you should aim to hold a particular speed on the airspeed indicator (ASI). Also try to keep the airplane on the runway heading, correcting if necessary for a crosswind. Your instructor will tell you at what speed he wants you to fly; probably it will be about 75 mph.*

Depending on your airfield, at about 400 feet above ground level (AGL) you'll probably start a climbing turn to your left to a pat-

---

* In a Grumann American Trainer it would be 95 mph.

tern altitude of 800 feet. This varies from airport to airport, depending on ground obstructions, homes, and industrial development.

During takeoff, if there is a crosswind, you will quickly discover a tendency for the aircraft to drift off course. Possibly you will also notice a tendency to steer left as a result of the spiraling airstream from the propeller hitting the vertical fin. In most modern trainers this is hardly an appreciable factor, and a very light touch of rudder at full throttle is sufficient to counteract this. For these two reasons, one of the very last checks we made before opening the throttle when the airplane was lined up, was to check the DG in relation to the compass and runway heading (pp. 23–24).

Runway headings are magnetic (don't forget local variation) and are used in this last-minute check. Now, having confirmed the DG, you simply note your heading from time to time as you ascend. If you find you are drifting off course, correct until you pick it up again. This is one of the things it is difficult to learn from a book, but you'll learn it quickly and easily with practice. When you're proficient, even in a heavy wind, you should be able to stay with the runway center line until ready to start your climbing turn.

Because flaps lowered in small degrees can be helpful in providing additional lift, shouldn't you use flaps to get airborne at a lower speed? This depends on the airplane. In the Cessna 150 Trainer (Figure 3.5), for example, 10 degrees of flap will shorten the ground run by approximately 75 feet, but this advantage is lost during climbout to 50 feet, the height of the "usual" obstacle at the runway's end. So you'd use 10 degrees of flap for takeoffs only from a soft field where there were no obstacles, and you would not retract them until you had cleared the "runway." During high-altitude takeoffs the use of flaps is not recommended. (Check the use of flaps with your instructor and in the Owner's Manual.)

With low-wing aircraft, partial and sometimes full flap can be used to reduce ground run on soft fields. However, the flaps might be damaged by stones and mud thrown up by the wheels during the takeoff run.

The Cherokee 140 makes use of as much as 25 degrees of flap for takeoffs from both short and soft fields. The technique here is to accelerate to about 55 or 60 mph and raise the nosewheel. Immediately on breaking ground you let the airplane accelerate in ground effect to its best angle of climb speed—78 mph—then clear what obstacles there may be, and finally climb out at about 90 mph. Where there is no obstacle, the technique is the same, except that on breaking ground you let the airspeed build up to the best *rate* of climb speed, 89 mph.

FIG. 3.5

Cessna Model 150

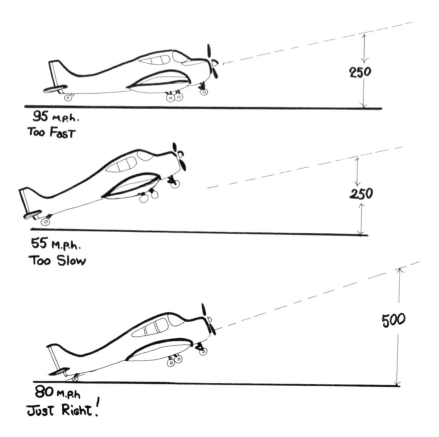

95 m.p.h.
Too Fast

250

55 m.p.h.
Too Slow

250

80 m.p.h
Just Right!

500

FIG. 3.6

Flyers like to achieve a flying altitude as efficiently as possible. Best rate of climb means proper engine cooling plus speedy altitude gain.

Although the best *angle* of climb speed will get you over your obstacle sooner, you actually have less room for error because you are closer to the stall speed. It is also much harder on the engine, as engine cooling at slow speed—with full throttle—is not as efficient. Therefore, unless you have a serious and genuine obstacle ahead, it is better to settle for the higher airspeed. If you are in any doubt at all, get a more experienced pilot to get the airplane out for you or get a truck.

The point is that with flaps your airplane becomes airborne at a lower airspeed; consequently the ground run is shorter. Although the rate of climb may be less, your forward speed is lower, and therefore, after approximately 40 seconds, you will have traveled a proportionately shorter distance, exchanging forward for upward motion. Figure 3.6 shows what happens.

ENGINE FAILURE

An interesting situation connected with takeoff is engine failure. If your pretakeoff checks are thorough, the chances of an engine failure at or just after takeoff are rather rare. Almost always such an occurrence is the result of something the pilot failed to do. The most common cause is water in the fuel system, with carburetor icing a close second. Remember that carburetor ice can happen even on a warm day. If you don't warm the engine during your taxiing and forget to do the carburetor heat check as part of your pretakeoff engine runup, it could happen to you. If something does go wrong, there are usually many earlier clues. The rpm drop may be excessive on one or both magnetos, in which case turn around and find out what's wrong. You don't need to be a hero by playing it cool; you are more of a hero by admitting that something could be wrong. When in doubt, always check it out thoroughly.

There *is* that million-to-one possibility that something may go wrong, and when it does, when your ounce of prevention doesn't work, it's good to have a procedure to fall back on. Your principal consideration will be altitude, as the more you have the greater your options. But remember that the two most useless things to a pilot are the runway behind him and the skies above him. For this reason don't make unnecessary intersection takeoffs—by taking off from the far end of the runway you might even get back and land safely on the remainder of a longish runway. Though on a shorter runway—3,000 ft. or less—this would be stupid—ask your instructor to show you why. Remember too that it is much better to run into the far hedge at 20 mph than the near one at 80 mph.

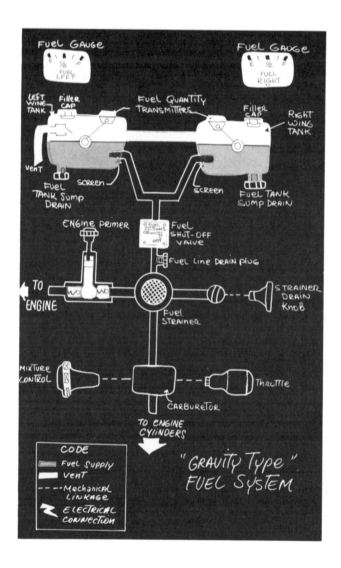

Because your course of action will be predicated on your altitude at that moment of mishap, you must bear one other factor in mind. The biggest pilot trap in the world *is* the temptation to try to turn around and glide back to the field. Unless you have plenty of altitude, forget it. Depending on your airplane, you will need at least 750 feet above ground level to turn around. It is dangerous, and you will probably be setting up the odds in favor of your becoming yet another victim to the stall/spin syndrome. Which is why it is suggested you get your instructor to show you—preferably on a mock runway at least 2,500 ft. AGL.

Follow this technique: Immediately lower the nose of the airplane to get your best gliding speed, while making gentle turns to avoid obstacles. You will try to deduce what is wrong by verifying that fuel switch, boost pump (if there is one), throttle, and carburetor heat are correct. At the same time your eyes will be searching for the best open space ahead in which to place your aircraft. Once you've spotted a good place, get set for that landing. In an emergency the time you have to spare is the time you create for yourself. Assured of your landing spot, use flaps or sideslip as needed to get in at a low landing speed. Once committed to your landing, switch off the ignition and fuel, and open the door.

What occurs more often on these occasions is that the pilot generally has sufficient power to hold altitude, in which case getting back to the field is not so difficult. If in doubt, follow the technique prescribed above. And do check your carburetor heat before takeoff.

One other item. When something does go wrong after takeoff, it most frequently occurs when power is reduced to cruise-climb; for this reason, it is sensible to maintain full takeoff power until you have sufficient altitude in hand to go back to the field if necessary. (This is especially true of failures with prop governor systems in aircraft with variable pitch or constant-speed propellers.) Your instructor will practice this important drill with you.

Power-off Emergency

Another item you'll come to by and by in your emergency procedures program is the simulated engine-out. This is an unpowered landing. You will be nervous the first time your instructor tells you he's going to show you how it works, and the resulting silence can be scary. But don't worry about it. Even without the engine, an airplane will glide, although not as well as its unpowered cousins, the gliders. Simply establish the best glide speed, find a place to put the airplane down, and get to it.

It must also be remembered that fuel gauges are not as reliable as they might be, and some aircraft do perform less well than the Owner's Manual indicates, especially if the engines are out of tune. For this reason, the most reliable fuel gauge is the clock on the instrument panel or the watch on your wrist.

If you made a visual check of the fuel level before you took off, and if you know what the consumption of fuel per hour is and the capacity of the tanks, you'll have a very good idea of how long you can stay up. If for some reason you have miscalculated your fuel, by reason of a headwind, for example, and you are in danger of running low, a precautionary landing with power is a lot easier to handle than a powerless landing. Not that the landing without power is difficult, but you only get to do it once, whereas landings with power leave you plenty of options.

In flight experienced pilots always know where the nearest airport is along their route, and when they are some distance from the nearest airport—if not within gliding range—they automatically check the countryside below for potential landing areas. It is good practice, therefore, when you are planning any cross-country flights, to make a mental note of those airports that are near your route and to note the topographical features along the way.

Night Visual Flight Rules (VFR) pilots frequently do not follow direct routes but travel slightly farther by flying a course from the environs of one airfield to another along their way; this is a good idea. In the northeastern United States it is very difficult to be more than twenty-five miles from some sort of field in which you could put down, and at night you certainly enhance your safety factor by following this method.

The principal problem with landings without power is that there is not much time to decide which is the best place to put down and little time to inspect the actual surface. Finding an area long enough for landing and subsequent takeoff is not usually difficult—in an emergency, even a ploughed field can be used, provided you land in the furrows. But ploughed fields are not the best landing places—unless there are no others—because the ground is frequently soft or soggy. A grassy pasture is ideal, and is usually found relatively close to a road or farm.*

At the same time that you're selecting the field in which to land, you should also be making a final check on the wind direction. You will presumably remember the wind direction at the point of your departure, and during your navigation you will have noted in which direction—if any—the wind has changed. Now you make a final note to confirm your earlier observations. Because the wind aloft may vary somewhat compared to the wind on the ground (though usually not by too much), it is worth checking below for clues. For example, you can sometimes see smoke from chimneys or the movement of trees, leaves, boughs, and shrubs—all are helpful indicators. If there's a lake, the effect of the wind blowing across the water can be an excellent measure of which way the wind is blowing and how strongly.

* "If the pasture weren't full of rocks it would be ploughed."—Old Texas Proverb.

There now comes a point of much importance. In emergencies either a left- or right-hand pattern applies—the choice is yours. The keypoint to your landing, however, is an imaginary mark placed at the end of the theoretical downwind leg to your landing field. You should try to arrive at this point at about 1,000 feet AGL, for what follows depends on having this amount of altitude above the ground. (We are speaking here about the ideal situation.)

Now, having selected the field in which you will land and having planned the circuit that will take you to it, you can get on with finding out what has gone wrong. Fuel and carburetor heat are the first two items. Check fuel, check mixture control, try switching tanks if you have more than one, and switch on the fuel pump, if there is one. The chances are quite good that carburetor heat may solve your problem. If not, proceed with the emergency landing.

We had settled on a key point that was to be 1,000 feet above ground level and situated at the end of a theoretical downwind leg to the field you have chosen. We have now arrived at this point and still can't get the engine to work, so we commit ourselves to landing. All the downwind checks should be completed, and fuel and ignition should now be switched off. Seat belts should be fastened securely and tightly. Some theorists hold that the door—if there is only one—should be wedged open.

Now, at the key point one of the things you must do is to plan to overshoot the landing. It gives you more altitude to play with your glide control. You can always lose altitude by adding flaps or by sideslipping, but without engine power you cannot climb. Also, it is much better for you to run into the field's farther boundary at taxi speed than it is to fly into a fence or hedge at the beginning of your emergency strip.

At the key point, if you are not too busy to forget it, you can also put out a Mayday call. If you are in contact or have been in contact with a ground station recently, it would make sense to call on that frequency first. Otherwise, call on the International Distress Frequency at 121.5 MHz. (You will find how to make an emergency call in Chapter 10.)

Now make a gliding turn onto your base leg. You will be able to check the wind by the amount of your drift from the field; if it seems excessive, correct it by turning slightly inward to the strip. Your base leg should be quite close to the field; Figure 3.7 shows the various methods you may use to lengthen your approach in order to lose excess altitude or to correct for not having sufficient altitude.

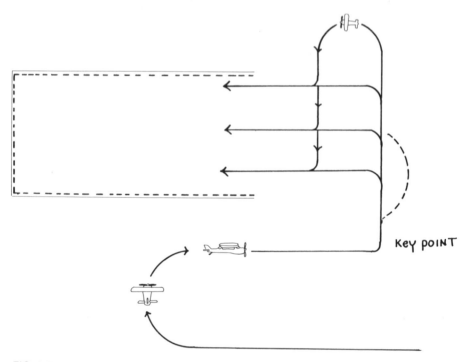

FIG. 3.7

Emergency landing technique

The advantages of choosing the key-point technique in making emergency landings are two: Having a practiced procedure to follow reduces stress, and it gives you a number of options as to where to turn onto final approach. It's not perfect, but it is a lot better than just guessing; if you get to your key point with some excess altitude, you can sideslip away your additional height, or if you're a bit low in altitude you can turn into the field almost immediately. If you have plenty of altitude you can extend your base leg to the other boundary of your emergency strip and make a 180-degree turn toward the field to run a second base leg. Incidentally, never turn away from the field you've selected for your emergency site, as you will find that you lose more altitude than you expect in the turn. Anyway, as a last resort you most likely will be able to use the flaps.

If you have used the flaps, switch off the master switch at this point. (If your flaps are not electrically operated, turn off the master switch at the key point.) In practicing such an emergency it is not usual to land on the practice field, though you may be given a few test runs at your own airport. Engine cooling occurs quite rapidly when the throttle is cut, and it is usual to give a little burst every 500 feet or so on the way down to prevent the spark plugs from oiling and the engine from overcooling. You will, of course, be using carburetor heat when you throttle back; otherwise you might convert a practice run into reality.

Power-on Emergency

The objective of the power-on emergency landing is to give you a chance to land before all your alternatives have run out. You will select the site in much the same way as before, except you now have an opportunity to check out the lay of the land. An initial dummy run is made at about seventy-five feet above your landing area, checking for ditches, animals, shrubs, and potholes. Before descending to this low altitude, you should make a very careful check for power and telephone cables. You should also check for barbed wire fencing around your landing zone.

A useful tip before commencing with ground inspection is to note the approximate heading of the field so that you can keep your circuit relatively close to the field. Make your turns 90 degrees each rather than 180 at the far end of the field to bring you back to the key point. Since you want to be ready to touch down almost as you come over the field's boundary, it's helpful to extend the key point slightly farther downwind than when making an emergency landing without power. If there's a fence at the near end of the field, make a steep descent using power to drag the airplane in with flaps and cut the power as soon as you see you have it made. Very short landings can be made with this technique. As your rate of sink may be high, you can check it with a touch of power as you break the descent in the flare—but practice this with your instructor, having him first show you how it's done. In both instances remember that pointing the nose down more steeply helps stretch your glide, whereas bringing the nose up steepens your descent.

In both instances once you have touched down, brake as quickly as is reasonable. If the emergency was a power-on landing, switch off the engine and inspect the ground before taxiing the airplane. If you have to leave the airplane in the field, secure it and attempt to find the owner of the field to let him know what is wrong and why you're on his property. Most farmers are surprisingly understanding and will usually try to assist you in putting matters right. (Unless you hit one of their cows.)

CARBURETOR ICE AND OTHER PROBLEMS

Carburetor icing is a potential hazard under almost any conditions when the outside air temperature is between 0° Fahrenheit up to about 80° Fahrenheit, and it does not have to happen in cloud. If you are experiencing carburetor icing, your first intimation that all is not well will be a subtle drop in engine rpm. This may happen quite slowly. However, if you're flying through a snowstorm, or even a heavy rain, it can happen quickly.

If you're not aware that carburetor icing might be the cause of the drop in rpm, you might first of all think that the throttle friction nut had worked loose and simply apply more throttle to correct the drop. If you do that, you'll experience a further drop in engine rpm and the onset of roughness. If you still don't do anything, the engine will quit, sputtering to stay alive—but it will ultimately quit.

Obviously you don't want that to happen. You therefore apply carburetor heat. Quite frequently this seems to make matters even worse. Don't worry if the engine sputters, gasps, and belches as it tries to digest the pieces of ice it is swallowing. It may seem about to expire. But then, at the very edges of eternity, the engine catches again and all is well.

The factor involved in carburetor icing is the relative humidity of the air in relation to the temperature. As a good rule of thumb, if the dew point is within about 5° of the current air temperature, there's a better-than-average chance of experiencing carburetor icing. The reason for this is that when air is brought into the carburetor, it goes through a venturi mixing with the gas. As the air moves through the venturi, its temperature drops; if the dew point is close to this newly acquired temperature, ice can form, depending upon the relative humidity. Ice can also form because of the evaporation of fuel; once again, the means required for this evaporation comes from the incoming air, cooling the fuel to a point at which ice may form.

The ice formed upsets the fuel/air ratio of the mixture going into the cylinders; at the same time it tends to increase the formation of more ice, until finally nothing gets through the venturi. The carburetor heat system is designed to bypass the inflow of cold outside air that is normally fed to the engine and to bring warm air from the engine bay. This will melt the ice if used in time. However, there are problems involved in using carburetor heat. One is that with warm air the engine produces less power (the principle of the internal combustion engine is that of taking cold air and expanding its mass by heating). The other problem with using carburetor heat on the ground is that this air is unfiltered, and frequently dust particles from the ground can be introduced into the cylinders via the carburetor heat air supply system. Except for the pretakeoff check, therefore, carburetor heat is not normally used before take-off unless absolutely needed.

Another cause of carburetor ice is the same as that of airframe icing—cold air and a high dewpoint. If there is even the slightest buildup of ice on any part of the fuselage or wings, you should watch out for ice in the engine too. The simplest method of treating the internal icing problem is by using fuel injection. Most expensive aircraft today are fitted with this system. Instead of using the carburetor, a direct fuel injection unit is used to measure the correct quantity of fuel and air for the local atmospheric temperature and pressure. A separate nozzle introduces this mix into each combustion chamber; because each cycle is very accurately metered, fuel injection offers several economies over normally aspirated engines.

A NOTE ON LOW FLYING

Generally speaking, all pilots should avoid flying low. There are too many obstacles around—television pylons and various towers and wires. Students and low-time pilots who wish to discover the problems involved should always make their earliest attempts with an instructor. Although low-level flight is an interesting exercise, it is a cause of considerable nuisance to those on the ground and is dangerous for the inexperienced.

In addition to the handling of an aircraft near the ground, there is also the problem of navigation. For a start it's sensible to complete all the cockpit work ahead of time, since you will be occupied in keeping the airplane at an appropriate altitude. Map-reading and course-keeping become important, as you don't have much time for verifying checkpoints and so on. High tension electricity cables are almost invisible, as are television towers in poor visibility, until you are almost upon them. Finally, there is the problem of maintaining a reasonable amount of altitude. This can be done only by constantly checking the ground ahead and below. And if you are over water you had better fly on instruments, since you can quickly come to grief over water that is smooth. The same applies to low-level flight over snow. In poor visibility the dangers are increased considerably.

Hedge-hopping is really for the military and is intended to avoid an enemy's radar cover. The only time low flying should be attempted by the ordinary pilot is to check his emergency landing strip, unless the pilot is a crop duster or involved in aerial surveying or photography. It is dangerous, and each year there are a number of accidents that could have been avoided if the pilot hadn't given in to the whim to show off to his friends.

There are four principal factors in low-level flight. First, especially in summer, uneven ground heating can make for downdrafts. The hand should be kept on the throttle at all times during low-level flight. Second, violent changes in attitude occasioned by an overly quick maneuver can lead to sink or even to a high-speed stall—for example, pulling up suddenly to avoid an obstacle on the ground. Third, communications are likely to be poor close to the ground, and you may find that the Very High Frequency Omni Directional Range (VOR) signal goes dead unless you are quite close to the station. Fourth, the effect of the wind on the airplane's path is highly apparent and can lead to confusion on the part of the pilot, especially when flying downwind. Keep a close eye on the airspeed indicator, since it is airspeed—not groundspeed—that keeps the airplane up. Particularly dangerous is the optical illusion that occurs during a turn. A turn into a headwind from a downwind flight path will make it appear that the airplane is skidding outward, while a turn downwind after flying into the wind will create the impression of a sideslip and that the airplane is speeding up. A constant check on the instrument panel can help to alleviate this problem, and the pilot should not make corrections with rudder until he is sure that the condition is for real.

Let's go a little higher so we can take a look at some ground-reference maneuvers. The object of learning how to make these maneuvers is to gain confidence in handling the airplane near the ground and to learn to handle the aircraft naturally when posed with the problem of joining the circuit at some of the more unusual fields, for example, where the pattern to be flown is triangular, with mountains on one side, a forest on a second side, and a deep gulley on a third. The object of the maneuver is to cover a prearranged pattern over the ground, correcting for the effect of wind both in straight flight and in turns. Needless to say, these maneuvers are best practiced when there is a reasonable amount of wind—and when it's a little bumpy.

The secret of good performance is to know what you are going to

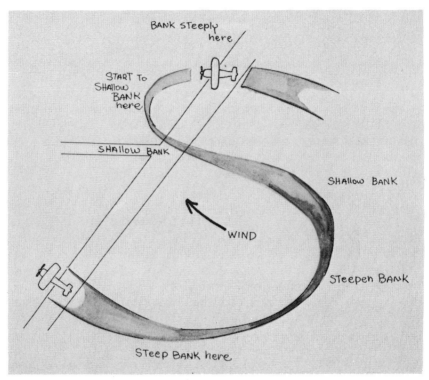

FIG. 3.8

S-turns across a road

do ahead of time. This may seem obvious, but if you have a firm idea in your mind of the result you wish to achieve and the methods required to achieve it, you are more than halfway there. For example, if you think about how you will redefine your path if the wind should exert a stronger force than you anticipated, you are ready to succeed before you begin.

There are a number of useful ground reference maneuvers to sharpen your skill, such as S-turns across a road or railroad track, or figure-eights, or constant turns about a point. For the S-turn (See Figure 3.8) an altitude of not less than 1000 feet above ground level (AGL) is recommended, preferably away from any habitation. The reference line—railroad track, road, whatever—is now crossed at right angles downwind. Since the speed will be highest with a tailwind at the actual point of crossing on the downwind approach, a steep bank is necessary to compensate for the speed, which is paid off little by little during the turn until the reference line is reached once more with the airplane flying into the wind.

74

At this point a turn in the opposite direction is initiated. First begin with a very shallow turn, which little by little is steepened until the reference line is crossed once again in a downwind direction. Then start the whole procedure again with another steep turn in the direction of the original turn, and so on. The object of this exercise is to inscribe the sky with equal semicircles about the reference line. In a strong wind this is sometimes easier said than done.

Figure-eights (Figure 3.9) are similar to S-turns in that they simply complete each circle at the reference point; they are similar to what are called pylon-eights. The 720-degree turn about a point is a steep version of an ordinary turn about a point. Again the object is to select a point and to maintain a constant radius from the point chosen at a constant altitude. It is quite easy to get hypnotized by the point in question, so a close watch must also be kept on the altimeter and for other aircraft.

Before starting these turns about a point, make a mental note of four key objects that are approximately equidistant from the point selected so that you can visualize a ground path to follow. You will naturally be making corrections in your angle of bank, depending on whether you are being affected on the one side or the other by the wind. You should also watch your turn and bank coordinator, as the maneuvers are supposed to be precise.

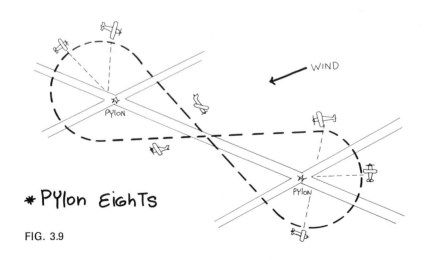

*Pylon EighTs

FIG. 3.9

THE STEEP TURN

The steep turn is merely a developed medium turn. When we discussed spins, it was suggested that an instructor demonstrate the spin that can arise from a turn. It is actually in the steep turn that the danger of a spin is increased. In the steep turn the stalling speed of the airplane rises in miles per hour, as the angle of attack of the airplane has to be increased in order to compensate for forces that are developed during the turn. The point is that if you steepen the angle of bank, you also automatically increase the loading factor of the wing.

When we described the forces acting on an airplane in level flight, it was pointed out that in undisturbed air we balance weight against lift and drag against thrust. This is to say that the load factor is 1—the wings are supporting only the weight of the airplane. If you bank the airplane quite shallowly in a 20-degree turn, the wings now have to support a load factor of 1.065. Double your bank to 40 degrees, and it goes up to 1.31. Add a further 20 degrees to bring it to a total of 60 degrees and the load factor becomes 2. A further 10 degrees brings the load factor to 3. From this point on, the load factor advances to a factor of 10 at a bank angle of 84.3 degrees. Most light aircraft just don't have sufficient power to turn so steeply.

To provide the lift required to balance out these loads we have to increase the angle of attack. We'll also need every bit of help we can get from the engine for additional thrust. Again, as the stalling speed of our airplane increases in a direct ratio with the square root of the load factor, it is sensible to be prudent about where and at what speed you make your steep turns. Steep turns at slow speed are not merely stupid, they're suicidal. A good rule of thumb to apply to light aircraft when making steep turns is not to exceed more than about 60 degrees of bank, since the engine power available is usually insufficient to overcome drag without losing height.* (See Figure 3.10.) Most instructors will demonstrate this to you, but you will probably be asked not to make turns steeper than 45 degrees.

In steep turns maintain altitude. Don't be surprised if in your first attempt you lose altitude. There is a simple technique that can help (and it is much easier if you are in an aircraft with a control stick rather than a control wheel): Roll in enough bank for your

---

* Over 60 degrees is aerobatic—illegal without chutes.

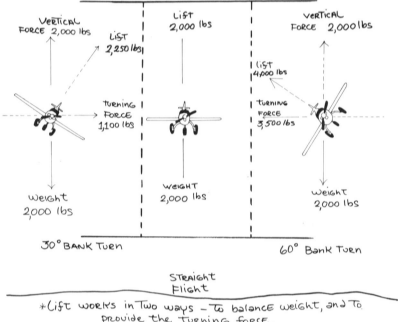

VeRTICAL FORCE 2,000 lbs
LIST 2,250 lbs
turning FORCE 1,100 lbs
WEIGHT 2,000 lbs
30° BANK TURN

LIST 2,000 lbs
WEIGHT 2,000 lbs

VERTICAL FORCE 2,000 lbs
LIST 4,000 lbs
turning FORCE 3,500 lbs
WEIGHT 2,000 lbs
60° Bank TURN

STRAIGHT Flight

*LIFT works in Two ways – To balance weight, and To provide the turning force.

FIG. 3.10

turn and as it establishes itself, apply back pressure on the control wheel to keep the nose up. You should be able to move the nose across the horizon in a straight line. In the meantime your other hand has been adding power: more power for more bank is the rule. Steep turns are best gauged *by the attitude of the nose in relation to the horizon*, and you will find it quite simple to select a point and hold the nose in reference to that point.

It's important to learn the coordination of these control inputs; for example, if you apply back pressure too soon, you'll start to climb, and if you apply it late, you'll already be going down. Similarly, once the nose has come through the horizon—unless you've done it properly—there is a serious tendency to begin a spiral. Adding more back pressure is not going to help you until you ease your angle of bank, so any tendency for the nose to drop should be corrected instantly.

You can do a steep turn in a glide, but remember to increase airspeed above what you would normally use for a normal gliding turn. The stall speed of an airplane increases dramatically as the bank angle goes above 60 degrees.

One last general rule about turns: If you're slipping or losing height, the chances are that you're not feeding in sufficient back pressure to the control column in relation to your angle of bank.

77

4

Learning on the Ground

This chapter deals with what you will learn at ground school. It includes weather, the basics of air navigation, air traffic control, the federal aviation regulations as they apply to pilots, the Airman's Information Manual, and a consideration of the internal combustion engine and its distant cousin, the gas turbine.

WEATHER

Weather is one of the more neglected aspects of pilot training, which is a pity since it is a fascinating subject. Aside from being an intriguing hobby, weather is vital to the pilot—for it is always changing. And because current plans call for an automated weather information service in the early 1980s, pilots are going to have to become even more expert in weather interpretation.

Weather is a related factor in about one-fourth of all general aviation accidents. Of course, it's the pilot who is actually responsible, as around 70 percent of these accidents are of the stall/spin variety. The pilots went beyond their limits, but weather was a related cause. Still, the more we know about how weather can be a limiting factor to our abilities, the better we can learn how to make accurate predictions of our own—either based on data supplied to us from television, the radio, or the weather service, or from a blend of these observations and our own local knowledge. Indeed, our own interpretation of weather can be a very real key to our own safety.

One of the best maps available to the pilot who wants up-to-date weather information is the one that appears early each morning on

NBC's *Today* program. Prepared by a meteorologist just before the show goes on the air, it's similar to the weather maps you'll find at your friendly Flight Service Station (FSS), and is a very good place to start deciding whether or not you'll be flying that day. In any case, you'll call the FSS.

Obviously, the more we know about the subject of weather, the better our interpretation of the weather maps is going to be. First, we need to know that air is not stationary; it moves around and adopts the characteristics of the region from which it is moving. Air coming from the poles will be cold and dry; air moving off the Caribbean Sea will be moist and warm. A front is the line at which two different air masses meet.

If we go deeper into the subject we discover a number of factors bearing on how weather is made. We might, indeed, start with the point that the earth is spherical in shape, and that it is a planet that rotates on an inclined axis and revolves in orbit around a star we call the sun. This provides us with day and night, and with our various seasons.

Around the planet, clothing it from the cold emptiness of space, is a mantle of gases we call the atmosphere, divided into three basic layers. The outer layer is called the ionosphere, made up of very finely charged molecules. These don't appear to have much to do with weather—our weather takes place lower down—but this layer is important to radio communications. (Between the ionosphere and the stratosphere lie the exosphere, thermosphere, and mesosphere, which we don't need to deal with here.)

The next layer is the stratosphere, which starts at 35,000 feet and extends up to about 30 miles. The higher you go into the stratosphere, the greater the heat caused by the absorption of ultraviolet radiation from the sun. This radiation is lethal; fortunately for us it is absorbed by the stratospheric ozone layer—the same layer that environmentalists fear may break down if SSTs become as common as jumbo jets. Normal jet aircraft use the bottom level of the stratosphere, where the temperature is usually $-40°$ to $-50°$ C. At the outer level it has risen to about $+10°$ C. because of ultraviolet radiation.

The third level is the troposphere, which contains about three-quarters of the total mass of the world's atmoshpere. It goes from sea level up to 35,000 feet. This is where weather really occurs. Wind, turbulence, sunshine, rain, hail, sleet, and snow are produced here by two principal factors—the action of the sun and the rotation of the earth. These forces also determine the moisture content of the atmosphere, moisture content being controlled by temperature.

79

The temperature of the atmosphere does not come directly from the sun. The sun's rays cut down through the atmosphere and are then reflected back by the earth's surface. This produces variations in local area temperature; for example, less heat is returned from a watery environment than from a desert.

Because of our planet's shape and tilt, the sun's rays don't strike the surface equally. At the poles they come down at an oblique angle, and consequently not much heat is available for absorption and reflection; hence the polar regions are pretty chilly. At the equator, on the other hand, the sun's rays strike down on the earth's surface at an angle of 90 degrees, and reradiation is greatest here. As a general rule, the farther north and south of the equator, the cooler the temperature. There are exceptions—New York City in July and August is usually hotter than the Caribbean islands. Why? The cooling effect of the oceans and trade winds around the islands, and the influence of the land mass of North America on the latter. The near permanent Bermuda high during the summer sweeps moist warm air up with the south-westerlies.

If we were to trace this unequal heating of our atmosphere without regard to rotational forces, we would discover a cyclic effect. Air heated at the equator would rise, expanding as it got higher and losing some of its heat. Meanwhile, cooler air from the poles would flow into the equatorial area, replacing the rising air, to be heated and rise in its turn. The original heated air would now have cooled to the local high-altitude temperature, tending to drop downward toward the polar regions. Eventually it would work its way back to the equatorial regions.

However, because rotational forces are at work, this whole pattern of heating and cooling is modified. In the northern hemisphere the earth's rotation sets up a deflective motion of air toward the right; in the southern hemisphere this motion is to the left. The various oceans and seas give up moisture to the atmosphere and also cool the breezes on our shores. Thus, if you look at a typical weather system over the United States in summer or winter—when the weather is usually pretty stable—you'll find cool or cold dry air developing in the Canadian arctic moving south and southeast and warm tropical air moving northward from the central Caribbean area. The West is influenced by air from the Pacific—warm moist air in the south and cool moist air in the center and the north. Finally, in the East there is moisture from the Atlantic Ocean and the Gulf of Mexico.

How quickly each air mass replaces another is due in part to the rotation of the earth and to the temperature/moisture equation.

Professor Bjerkens, the Norwegian meteorologist, concludes that when a warm and cold air mass move adjacent to each other, a wedge of warm air intrudes into the cold air mass with a simultaneous drop of pressure. Added to which there is also the predictable cooling process of the air itself.

Our atmosphere is made of gases, which tend to heat up when compressed and to cool when permitted to expand. Air temperature drops at a rate of approximately 3.6° F. (2° C.) per 1,000 feet, with a further drop for expansion as the air rises, giving us an overall figure of about 5.5° F. (3° C.) per 1,000 feet upward. But air can be cooled indirectly by the moisture it carries, and there is always some moisture in the atmosphere, as sufficient cooling will make apparent in the form of rain, mist, fog, or even snow. The point is that the air can retain only a certain amount of moisture at any given temperature. This relationship between moisture and temperature is reflected in the dew point—the temperature at which the moisture begins to condense.

If you take another look at a typical weather forecast you'll find that both temperature and dew point are usually given, often together. We mentioned earlier that carburetor icing could be a problem when the air temperature and dew point are close together, but you could go further and say that most of our weather problems may be directly related to this phenomenon. Thus, the cooling of warm air not only makes for blustery frontal systems but also produces beautiful and not-so-beautiful clouds and is even responsible for those miserable low ceilings when it rains and never seems to want to stop.

Wind is simply air in motion and is caused by the movements of different air masses over the planet and their interaction with each other. The rotation of the earth produces reasonably predictable local systems, which are responsible for certain fixed wind patterns, such as the trade winds in the tropics, the northeast trade to about 30° N latitude, the prevailing westerlies to about 60° N, and at the top of the sphere the polar easterlies. Wind direction is always reported *from* the direction the wind is flowing.

Air near the ground moves around trees and hills in the countryside and around buildings in towns and cities. In large cities miniature weather systems form from time to time around tall buildings. On such days you'll expect to find turbulence up to a good height, and your ground speed is likely to be affected—for better or worse. Above 20,000 feet most winds are westerlies, which means that easterly winds tend to decrease with altitude. North winds back to the west, and the winds from the south tend to veer. When you're flying and there are clouds, a good way to gauge the real wind is to watch the cloud shadows moving across the ground.

Looking again at the weather map, we notice that there are numerous lines that look like gradient lines. These isobars are actually pressure gradient lines. When there are wide spaces between the lines, the gradient is small, and the winds tend to be light. Conversely, when the isobars are close together you may expect to find strong winds.

The pressure gradients are boundaries of high or low pressure systems. Air tends to move inward and counterclockwise around low pressure systems; the reverse is true with high pressure systems. In the southern hemisphere the direction of air movement is the reverse.

Upper winds flow parallel to the isobars; because of the friction between the air mass and the terrain, lower winds tend to flow *across* the bars. Over water, where there is relatively little friction, you'll find about 10° direction difference between upper and lower winds. But over the Rocky Mountains you might find from 30° to 50° direction difference between surface winds and winds 6,000 AGL. The greatest differences are near the ground, and this condition is lessened above 3,000 feet.

Another look at the isobars will show you the areas of frontal activity. This describes the meeting point of two air masses having different characteristics of temperature and moisture content. Frontal weather is the precursor of change, and when a front passes through you'll see a change in the weather.

Cold fronts are the most spectacular. As they move toward a warm air mass, the heavier air tries to push up the warm air and sneak in underneath. Cold fronts tend to be aggressive and are classified by their rate of movement. The fast-moving cold front is more common in winter than summer and can move across country at up to 45 mph. Within the front there may be winds of more than 60 mph, and a common feature is a squall line parallel to the front and running 20 to 200 miles ahead of it. You have probably seen a squall line come bursting in during the summer months, and from the air you can usually see fine weather following behind the squall. Frequently, squall lines may be associated with cumulus clouds.

Slow cold fronts can produce heavy cumulus cloud cover if the weather ahead is moist and unstable. If not, stratiform clouds will develop. In the summer a cold front can produce cumulonimbus clouds with thunderstorms. Eventually the cold front catches up with the warm front to produce an occluded front, in which case the weather is likely to consist of widespread rain and cloud cover, with moderate to poor visibility and the danger of icing in the clouds.

A warm front occurs when a cold air mass retreats before a warm air mass. Typically, the warm air rises in a slight incline over the cold air. The speed with which the front moves is around 15 mph or less. As the warm air rises, its moisture content cools and forms clouds and can in certain instances produce cirrus clouds at 30,000 feet and higher. Cirrus is a good indication of a warm front in the offing. Below the cirrus clouds are the cirrostratus, followed by the altostratus and then the rain-laden nimbostratus. Heavy rains are found up to 200 miles ahead of an actual ground front, whereas cirrus sometimes precede the warm front by as much as 600 miles. Such air masses can cover sizable areas of land or sea.

The trouble with warm fronts, as far as pilots are concerned, is the pockets of cumulonimbus found from time to time within the cloud mass itself. Other problems are low-level stratus and fog— both due to the high humidity of the atmosphere. If the cold air is considerably colder than the warm air that is displacing it, freezing rain and sleet can develop in addition to the usual inconveniences of poor visibility and fine rain.

There are two other types of front—the stationary and the occluded. The stationary front is, as its name implies, a front that isn't moving or is showing little movement. It is caused by the equal pressure of two adjacent air masses. The weather associated with a stationary front is similar to but usually much less intense than a warm front.

The occluded front occurs when a warm front is followed by a cold front, but both are subjected to their own centrifugal reaction, which causes their edges to flow outward. An occluded front can be either warm or cold, depending on what type of front develops at the edge of the low pressure system. Occluded fronts usually mean widespread rain and clouds with limited visibility and a very real danger of icing exists in most frontal systems depending upon the time of year.

Fog can be described as stratiform cloud that is within fifty feet of the ground. Humidity must be high, and the dew point must be close to the air temperature. If they are within 4° F. of one another, you can usually expect fog, unless some wind is predicted later. You can be certain of fog if the temperature spread is less than 2° F. A thick fog needs a very light surface wind to help it "brew." Without any wind, fog will simply exist in patches, usually in hollows close to the surface, as you may have seen on certain evenings when driving late. Fog will not form in strong winds.

Ground fog (radiation fog) is normally burned off by the sun as the temperature rises. It is caused by the cooling of the air near the surface of the earth and occurs when the skies are clear. The effect of the cooling is to bring the temperature and dew point close together, which is sufficient to produce the fog.

Advection fog occurs when warm humid air flows over a cold surface. You frequently find this in coastal regions where there is a difference between land and sea temperature. The warm air is cooled by the ground, and as soon as dew point is reached, fog is formed. Advection fog may contain winds of up to 12 knots, which help it to thicken; at winds over 15 knots, the fog will usually lift to form a low layer of stratus. Advection fogs, when they form over the sea and are blown on shore, are called sea fogs.

Another water fog—steam fog—occurs when cold air moves over water that is warmer than the air. The relatively intense evaporation from the water into the cold air causes the vapor to condense, forming fog. Fog may also be caused by excessive rain, or when rain evaporates on contact with the ground. This type of fog can get thicker at night. Finally, there are mountain and hill fogs, caused by the flow of moist air moving up the slope, whose moisture, with the temperature drop, condenses as it cools. This type of fog depends on a breeze to push the moist air continuously uphill.

Haze is caused by fine particles of dust suspended in air. Smoke has an effect on visibility similar to haze but is produced mostly from industrial furnaces. Smog originally was recognized as a mixture of smoke and fog, but today the word is more usually applied to any heavy concentration of air pollution. Another term you sometimes hear is smaze—smoke and haze—which also indicates restricted visibility.

Thunderstorms produce considerable turbulence and also ice. However, if you've done your homework properly—and with the availability of accurate weather data to pilots across the country it's difficult not to—you should be able to avoid such conditions quite easily.

Safe flying depends on getting the overall weather picture first and then the details. The overall picture will tell you what the current state of affairs is; by developing your own insight into weather behavior, you'll be able to predict what may happen over the next few hours. But you can make it more accurate than that by getting a full weather briefing before you leave. The more recent your weather information is, the more accurate your own forecast will be.

The National Weather Service is accurate at least 75 percent of the time in forecasting certain types of weather, such as the passage of warm fronts or slow-moving cold fronts up to twelve hours in advance, plus or minus five hours. Fast-moving cold fronts and squall lines can be predicted as much as ten hours in advance, plus or minus two hours. But forecasters still can't tell you where you'll find turbulence or icing, or where thunderstorms will develop before they exist; they can only suggest that one may be likely or probable.

How does the pilot go about improving his forecasting ability? There are several ways, but the first thing is to have a good idea of the overall picture and then build on that. The *Today* show is one good way to start, if you don't mind getting up that early in the morning. Another—which you can use any time of the day and night—is the Pilot's Automatic Telephone Weather Answering Service (PATWAS) or the Transcribed Weather Broadcast (TWEB) service in many parts of the country.

After you've heard the general area weather, you have a reasonably good idea of what to expect. But as the forecast is old—it was nearly an hour old when it was originally recorded—you need to

Weathermap: Typical of the weather map from which your flight forecast is prepared. Note the high pressure area over the New England states.

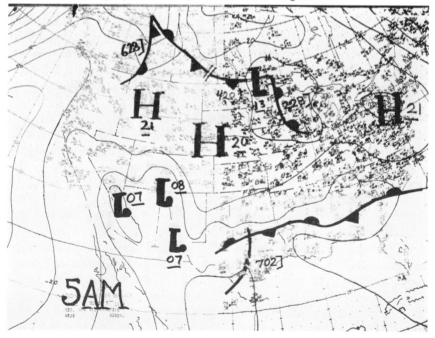

see whether everything is developing according to plan and whether there are some contingencies you should take into account. So now you call the local Flight Service Station (FSS). (Ideally, visit your local FSS before takeoff, in which case you may have access to satellite pictures and other data.)

Getting a proper weather briefing is nearly an art; you must ask the right questions to get the right answers. The weather-briefer has a gamut of information in front of him. Before you get a telephone weather report, you should have an area chart in front of you so that you can take notes. If you don't have one, make a rough sketch of the area and then note the salient points on it.

First, ask the weather-briefer to give you the synoptic weather— a summary of the weather at the last compilation. He will then get you current en route weather, together with terminal forecasts. He can give you winds aloft as forecast, and any Significant Meteorological Advisories (Sigmets), Advisories for Light Aircraft (Airmets), and Notices to Airmen (Notams).

Unless you are just flying locally—around the pea patch, as it's called—you should have a good idea of what your route is going to be. In fact, you should have tentatively planned two or three possible routes from your point of departure to your destination, just in case the direct route is weatherbound. (I'm assuming that you're flying VFR. If you haven't worked this out, then you'd be well advised to before you call or you'll be wasting the weather-briefer's time and yours.)

First, tell the weather briefer your points of departure and destination and the route you prefer. You should also tell him whether you want VFR or Instrument Flight Rules (IFR) and what type of airplane you'll be flying, together with the approximate flight time. He can then assess whether conditions are acceptable for your aircraft; for example, if your airplane is fitted with deicing equipment, it might be able to stand the light ice that has just been reported by a pilot at point X. The decision to go, or no-go is entirely yours, since *you* are the captain. You can expect the weather-briefer to give you the latest weather trends. Having called the PATWAS earlier, you now have an immediate picture of whether the forecast is holding true or is breaking down and changing.

In giving you this information, the weather-briefer uses a surface analysis chart, a weather depiction chart, and a radar summary chart. He may also use a satellite chart if a recent one is available. The surface chart depicts the highs and lows in the area and marks out the fronts. The weather depiction chart shows the local area as part of the overall weather system, so that it can be deduced how stable the forecast is likely to be. This second chart indicates those areas that are below VFR minimums or are marginal, and if you're a student pilot getting a weather briefing, it may be helpful to identify yourself as such. The radar chart shows areas of medium to heavy rains and any storm cell echoes (indicating thunderstorm activity).

The en route weather will contain a full sequence for your departure point, plus visibility and ceilings en route, unless you actually specify something different. The next item will be the terminal forecasts. Usually a weather-briefer—if you tell him at what time you're planning to leave—will give you a forecast for places along your route, along with that of your destination. If you were flying IFR, you would want to check with the weatherman for an alternate airport in case conditions at your destination airport fell below acceptable minimums while you were en route.

Knowing the velocity and direction of winds aloft helps in determining your groundspeed and what corrections you'll expect to make to stay on course. Wind forecasts are normally given up to 9,000 feet; you must request higher altitude forecasts. You'll be told if there are any Sigmets, Airmets, and Notams applicable.*

Area forecasts are twelve-hour forecasts with a further twelve-hour outlook of cloud, weather, and frontal activity for a general area covering several states, amended by Sigmets and Airmets. Terminal forecasts are updated every hour. The information they contain is encoded in letters and symbols, the first three letters being an identifier code, which tells you the issuing station. If the report is a special—which means that the weather has changed dramatically—the letters SP with a time group are added to the three-letter code.

Following the three-letter code are details of sky and ceiling. The sky is described as clear when less than one-tenth is covered; scattered, one-tenth to six-tenths covered; broken, six-tenths to nine-tenths covered; and overcast, more than nine-tenths covered. Next comes an indication of the current visibility and weather, with a letter or letters identifying any visual obstructions. Sea level pressure is quoted in millibars, followed by the station's temperature and dew point. Next comes a notation of the wind, its direction and speed in knots. The current altimeter setting is quoted in inches of mercury. Both the sea level pressure and the altimeter readings omit the first figure.

---

* Summaries of Notams—Nosums—are on file at the briefing center.

Finally, there may or may not be some remarks. At certain airports you'll find runway visual range and coded pilot reports noted before the remarks. Runway visual range (noted as VR in sequence reports) is a measurement of visibility made by a machine called a transmissometer; it gives actual visibility near the touchdown point of the runway mentioned in the report, in hundreds of feet. It is usually based on a sighting of runway high intensity lights or on the visual contrast of other targets, at a point near the touchdown area.

Sigmets warn pilots of potentially hazardous weather; a Sigmet applies to all pilots of aircraft and normally includes details of thunderstorm activity, severe turbulence, severe hail, severe icing, squall lines, duststorms, hurricanes, and so on. Airmets are concerned with weather hazardous to light aircraft and include moderate icing, moderate turbulence, winds of 40 knots or more within 2,000 feet of the ground, and decreasing visibility and lowering ceilings, especially in or near mountainous regions. Notams contain information of interest to pilots, such as airports closed for repairs, radio beacons inoperative, and so on.

Pilot reports (Pireps) are another useful source of information. They are actually filed by pilots. Tops of cloud formations, notes about turbulence, even reports on icing are to be found here. The Pireps column depends on pilots who contribute to it, and every pilot is urged to report significant weather.

En route weather reports started in 1972 on the West Coast and are gradually becoming available across the country in the form of the en route weather advisory service (EWAS). A total of forty-four sites are planned by mid-1976 to provide nationwide service. A discrete weather frequency of 122.0 MHz enables pilots to get up-to-the-minute weather advisories and Pireps from trained weather-briefers, but the service is only for en route weather. The filing of flight plans and preflight weather briefings are not permitted on this frequency.

A last word about getting a weather briefing. Even if the sky is clear and visibility unlimited, it's still worth getting a weather briefing; as a pilot, it is your responsibility. The FARs (Pt. 91.5) state quite clearly that "Each pilot in command shall, before beginning a flight, familiarize himself with all available information concerning that flight." Weather is very much part of the information all pilots should have before taking off.

NAVIGATION

The reason pilots should become skilled in navigation is not to ensure that they don't get lost every time they leave their home fields but to realize the full benefit of the aircraft they are flying. If a person can take off from one place and find his way across totally unfamiliar ground to a place he's never been but wants to visit, that pilot must be able to read an air navigation chart and must have prepared the flight thoroughly. The advantage of thorough preplanning is that one can concentrate and enjoy the actual flight itself, without having to worry.

Most pilots make use of routes called airways that crisscross the country. The airways are merely lines between radio navigational beacons, and unfortunately they don't exist everywhere. Therefore, the pilot must learn, first of all, to navigate without them.

Although we won't be covering celestial navigation here, we can still take as a starting point the fact that the earth is spherical. For navigational purposes it is divided into areas by north-south lines called meridians (usually expressed as longitude) and by east-west lines called parallels of latitude. (See Figure 4.1.) The north-south

FIG. 4.1

Parallels of latitude and longitude

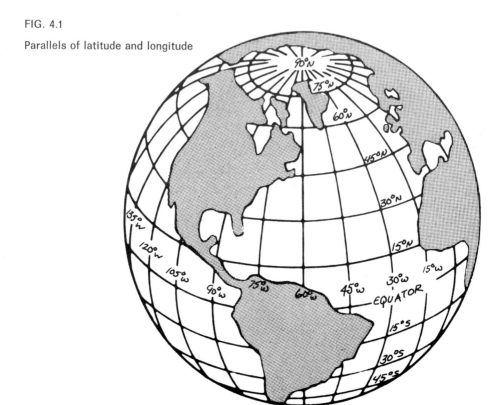

meridians are measured east or west of Greenwich (just outside London, England), where the prime meridian, 0°, is located.

The parallels of latitude are measured north and south of the equator. By using these referents, we can accurately pinpoint any place on earth. If we take our sectional charts—air navigation, like sea navigation, makes use of charts rather than maps—we would find Felts Field (about four miles east of Spokane, Washington) on the Spokane sectional at latitude 47° 41′ north, longitude 117° 19′ west. (The total circumference of the earth is 360°. Each degree is divided into sixty parts, called minutes, which are notated by the symbol ′.) That is to say, Felts Field lies 47° 41′ north of the equator and 117° 19′ west of the Greenwich meridian. John F. Kennedy International Airport in New York, on the other hand, is (on the New York sectional) at a latitude of 40° 38′ north and a longitude of 73° 46′ west.

Although this all sounds very simple, there is a slight problem involving the definition of "north." For although it is possible to depict a point as true or absolute north on the sphere that makes up our planet, the magnetic field that encompasses the earth chooses two rather different points for its north and south, and these magnetic points of north and south wander about from day to day and year to year. As a result, magnetic compasses become more eccentric the farther north or the farther south you go, and successful polar navigation uses different techniques. (See Figure 4.2.)

There is considerable evidence that the earth's magnetic field has reversed polarity many times (the latest evidence is 33 times in 80 million years). At present it is quite stable, and it is possible to predict fairly accurately the amount of wandering magnetic north will make in any year. However, at certain points local variations of up to 30 degrees, not always satisfactorily predictable, make themselves known. It is unlikely that you will encounter variations of more than about 15 degrees. But in the Vancouver area there are variations that may be as much as 25 degrees east, and in Newfoundland they may be as much as 30 degrees west.

Before making a journey anywhere, the pilot checks the local variation for the points along his route. (Magnetic variation is listed on sectional charts.) But there is a further problem—although the airplane compass will point to magnetic north if you take it out of the airplane and hold it in your hand, once installed in the aircraft the local magnetic field (induced by metal and electric wiring of the airplane) influences the accuracy of the compass. For this reason you'll find a deviation chart alongside the aircraft's compass that details the amount of error around the eight cardinal

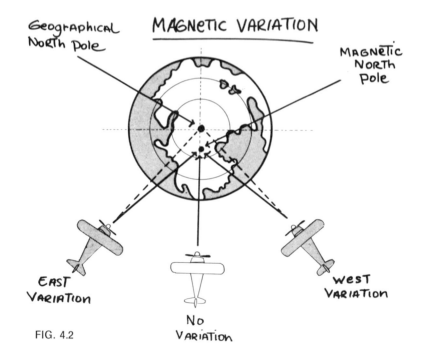

Geographical North Pole

MAGNETIC VARIATION

MAGNETIC NORTH Pole

EAST VARIATION

WEST VARIATION

FIG. 4.2

No VARIATION

points (north, south, east, west, northeast, southeast, southwest, and northwest). Since most modern compasses can be corrected by compensating magnets, the amount of error is usually quite small. But try to remember not to place a metal cigarette case alongside the compass.

The sectional chart that we've been talking about and that pilots use is a highly accurate map of a section of the earth's surface. The scale used is 1:500,000. In Britain this scale is referred to as the "half-million" series, and means approximately eight statute miles to the inch. (Nautical miles are longer—6,076.10 feet and 1/60th of one degree at the earth's equator. Statute miles are 5,280 feet long and derive from the Roman measure of 1,000 paces, which was approximately 1,620 yards. Presumably, taller soldiers take longer paces, which is why in the United States and Britain the statute mile now measures 1,760 yards.)

Sectionals are identified by the area they cover from north to south. The Detroit sectional, for example, includes the bottom half of Lake Huron, with the Toronto-Detroit area in its northern part; the southern side includes the lower coastline of Lake Erie through Toledo and Cleveland and extends across to Williamsport, Pennsylvania. Underneath the area name the type of chart (in this case, sectional), scale (1:500,000), and then details of the type of projection are noted.

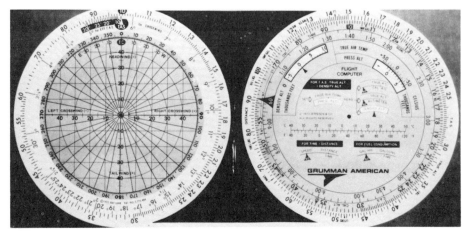

The CR series computer—ideal for solving problems

Projection is the process of representing a spherical surface on a flat surface. For the middle latitudes the Lambert Conformal Conic Projection is preferred, as a straight line on this type of projection represents very closely a great circle route, and scale inaccuracies are small. A great circle route is important for aircraft because the shortest distance between two points on a sphere will appear to be a great arc on a flat surface.

The data is followed by a date on which a new edition will be issued, thus when it will become obsolete for use in navigation. Although the cost of charts has increased immensely—without much reason one might add—it is vital to use up-to-date charts. Although television pylons don't grow overnight, it is quite possible that one could be erected and operational at a flight altitude you wanted to use during the six-months' lifetime of a chart.

PLANNING THE FLIGHT

The pilot marks a line on the chart between his departure and intended arrival point with a ruler and a soft pencil so that he can erase any errors or his marks when he has completed his flight. The pilot makes a note of the magnetic course, checking the local variation. Finally he checks the result against the deviation card by the airplane's compass. Although this sounds quite complicated, it really isn't.

The next thing is to select some checkpoints along the route. Because an airplane is not restricted by physical boundaries like a car and because there are no signposts in the skies, the pilot uses ground features that we call checkpoints. First he has to learn to read the chart, then he has to know how to follow his compass and his clock, and finally he must understand the effect of wind upon his flight. The checkpoints are used to confirm every so often that he is keeping to his intended course.

The distance between the checkpoints a pilot chooses will vary with the cruise speed of the airplane and with the pilot's confidence in his ability to fly a reasonably true course. Some people like to keep busy in the cockpit, and if the visibility isn't too good, this makes sense. These pilots select points about an inch apart on the chart, or every eight or nine statute miles. Others prefer checkpoints farther apart, perhaps two or three chart inches, or roughly every twenty miles or so. You should work out your own system and discover which suits you best on your cross-country flights. If you have to take quite a lot of wind into account, a check every ten miles or so is a sensible precaution until you learn how to correct for wind automatically. It may be worth noting, parenthetically, that as you become more expert in the art of flight, you will change your measuring of distance to nautical miles and your speed to knots. The world's air traffic control systems all use the nautical mile, which is one sixtieth of a degree.

Now that you've set up your route and selected the various checkpoints, the next thing you'll want to determine is how long the total journey time will be, and how long it will take from checkpoint to checkpoint. This brings us to a most useful device—the navigation computer, the best of which is the CR series, although most people still use the EGB. The navigation, or flight, computer has two sides. The side with numbers all over it is simply a circular slide rule, designed to help you work out problems dealing with time, speed, distance, fuel consumption, true airspeed, altitude, conversions from nautical miles to statute miles and vice versa, and most of the multiplication and division problems you'll need to solve. On the reverse side you may work out problems dealing with the wind triangle, or triangle of velocities as it is sometimes called in textbooks. The wind triangle deals with the speed and direction of the wind and its effect upon your heading, airspeed, course, and groundspeed.

The wind side of the navigation computer consists of a sliding grid, a true index, and an azimuth circle that rotates through 360 degrees. At the center of the azimuth circle is a transparent portion (you can write on it with a soft pencil) with a dot or grommet in the middle. It works as follows: You have already obtained the wind speed and its direction from the weather report, and you now set up the wind direction under the true index scale by rotating the azimuth circle. You next mark the wind velocity from the dot or grommet, merely by setting one of the thick mph lines directly underneath. If the wind is 20 mph, make a little x or dot at the point on the centerline showing an increment of 20 mph in relation to the figure under the dot.

The next step is to rotate the azimuth circle to put your true course under the true index. You will have figured this out from your chart. Now move the sliding grid so that the mark you made for wind velocity falls on the line of the airplane's true airspeed. Your groundspeed can now be read under the dot or grommet, and your wind correction angle is the amount of distance indicated between the centerline of the sliding grid and the wind velocity dot. Purists make a double check by feeding in the correction to the course, and seeing whether the wind velocity dot stays in approximately the same place—it should unless there's quite a correction to be made.

You now have your groundspeed and the course you will have to fly to maintain the heading you want, and you can now work out the time between those checkpoints. In flight you will check the actual time against your estimates to see whether any further revision may be needed. Unless the wind increases or decreases dramatically in force or direction (you'll have discovered whether there was any likelihood of this during the weather briefing) you'll probably stay with the estimated times of arrival (ETAs) you made. If you have filed a flight plan, however, it is good practice to update your final ETA with an en route Flight Service Station (FSS) if you might be more than fifteen minutes adrift.

An accurately prepared flight plan (whether you file it or not) is the best insurance against getting lost. (See Figure 4.3.) However, even the best of flight plans won't help if you can't read a map accurately. A useful tip is not to try to relate what you see on the ground to what the map says; see what the map says first and then relate it to the ground. Another point is to make sure of those checkpoints and note the time they come up, which they will do if you keep a close watch on your heading, and remember to check your Directional Gyro (DG) every ten or fifteen minutes. You can make a note of the time on the chart if you're too lazy to write down the route on a piece of paper—which is actually tidier and easier in the long run.

FEDERAL AVIATION AGENCY

FLIGHT PLAN

Form Approved.
Budget Bureau No. 04-R072.3

1. TYPE OF FLIGHT PLAN: FVFR / VFR / IFR / DVFR

2. AIRCRAFT IDENTIFICATION

3. AIRCRAFT TYPE/SPECIAL EQUIPMENT 1/

4. TRUE AIRSPEED — KNOTS

5. POINT OF DEPARTURE

6. DEPARTURE TIME — PROPOSED (Z) / ACTUAL (Z)

7. INITIAL CRUISING ALTITUDE

8. ROUTE OF FLIGHT

9. DESTINATION (*Name of airport and city*)

10. REMARKS

11. ESTIMATED TIME EN ROUTE — HOURS / MINUTES

12. FUEL ON BOARD — HOURS / MINUTES

13. ALTERNATE AIRPORT(S)

14. PILOT'S NAME

15. PILOT'S ADDRESS AND TELEPHONE NO. OR AIRCRAFT HOME BASE

16. NO. OF PERSONS ABOARD

17. COLOR OF AIRCRAFT

18. FLIGHT WATCH STATIONS

CLOSE FLIGHT PLAN UPON ARRIVAL

1/ SPECIAL EQUIPMENT SUFFIX
A — DME & 4096 Code transponder
B — DME & 64 Code transponder
D — DME
L — DME & transponder—no code
T — 64 Code transponder
U — 4096 Code transponder
X — Transponder—no code

FAA Form 7233—1 (4-66) FORMERLY FAA 398 0052-027-8000

PILOT'S PREFLIGHT CHECK LIST

DATE

WEATHER ADVISORIES	ALTERNATE WEATHER	NOTAMS
EN ROUTE WEATHER	FORECASTS	AIRSPACE RESTRICTIONS
DESTINATION WEATHER	WINDS ALOFT	MAPS

FLIGHT LOG

DEPARTURE POINT	VOR IDENT / FREQ	RADIAL TO / FROM	DISTANCE LEG / REMAINING	TIME PT-TO-PT / CUMULATIVE	TAKEOFF	GROUND SPEED
CHECK POINT					ETA / ATA	
DESTINATION			TOTAL			

POSITION REPORT: FVFR report hourly, IFR as required by ATC

ACFT IDENT	POSITION	TIME	ALT	IFR/VFR	EST NEXT FIX	NAME OF SUCCEEDING FIX	PIREPS

REPORT CONDITIONS ALOFT –
CLOUD TOPS, BASES, LAYERS, VISIBILITY, TURBULENCE, HAZE, ICE, THUNDERSTORMS

CLOSE FLIGHT PLAN UPON ARRIVAL

SCALE 1:1 000 000 — Nautical Miles / Statute Miles — WORLD AERONAUTICAL CHARTS

SCALE 1:500 000 — Nautical Miles / Statute Miles — SECTIONAL AERONAUTICAL CHARTS

FIG. 4.3

Flight plan (official form of the FAA)

During the early cross-country flights almost every student has a tendency to lose track of time—either because the flight is so exhilarating or because there seems to be too much to do. And so checkpoints occasionally get missed or don't appear on time, and the pilot gets nervous. Now, because the pilot is concentrating too much on finding that checkpoint, the heading may be forgotten— "Maybe I'd better fly over here, that looks as if it could be the checkpoint," says the tyro, and suddenly he has no idea where he is: The map doesn't correspond to the ground anymore.

Almost everyone gets lost sooner or later, so that in itself is nothing to worry about, provided there's plenty of fuel, and the weather is holding up. Unless the pilot has been highly inattentive, it is unlikely that he will be more than 10 percent of the total distance flown from where he actually ought to be. There are two helpful methods to ascertain your position. First, you can use the sun as a general guide: It will tend to be easterly during the morning until noon and westerly during the afternoon, which doesn't help too much, although it does save you 180° for a start. If it is shining and you have a wristwatch, you can use the old Boy Scout trick of pointing the hour hand to the sun, and splitting the angle between the hour hand and noon to give you north and south—yes, it does work, if your watch is reasonably accurate and you've become suspicious of the compass. Second, know how far you can see from a particular height: At 1,000 feet above ground level you can see almost 40 miles to the horizon; at 2,000 feet you can see nearly 55 miles; at 3,000 feet, 66 miles; and at 4,000 feet, 77 miles plus if the visibility is good.*

If you are lost, don't panic. First, note the time the last checkpoint was passed and the time at which you are going to start your "lost procedure." Check the remaining fuel and that you have sufficient altitude to fly over any neighboring high spots. If there isn't much fuel left, you'd better start thinking of an emergency landing with power. Meanwhile your DG should be checked against the compass, the error noted, and the DG reset. You are now ready to start your lost procedure.

You should have noted when you checked the DG what course you were on. If there was much discrepancy between the DG and the compass, the chances are that you may have become a victim of precession, in which case you will note that difference and take another look at your chart. Establish an area of uncertainty—a

---

* Actual figures based on a 4,000 mile radius: 1,000 ft. = 38.93 mi.; 2,000 = 55.02 mi.; 3,000 ft. = 67.42 mi.; 4,000 ft. = 77.85 mi.

square or circle will do—around which you think you ought to be in relation to the last checkpoint. If you use a circle, make the radius 10 percent of the distance you have flown; if you prefer to use a square, make half the diameter of the square equal to 10 percent of the distance you have flown.

You now have two options: You can either backtrack along your course until you come to the vicinity of your last noted checkpoint, or you can see if there is a road, railroad, canal, or river near you that looks worth following. (See Figure 4.4.) When you implement your decision, note the time—you are going to need to know this in any case because when you finally reach a checkpoint, you will need to reestablish that you have sufficient fuel to continue on your original flight. In addition, noting the time is a helpful reminder of how much fuel you will have used up until that point; also, you can calculate approximately how far out you might have traveled if you really have become hopelessly lost. (It is extremely difficult to become *that* lost if you've done any kind of halfway decent preflight planning. The biggest problems most usually occur in air pollution when you can't see anything very much from your safe altitude. It is in this case that it becomes very important to follow a proper lost procedure and stay with it. If you have plenty of fuel and night isn't pressing in, you have plenty of leeway. Above all, don't panic.)

If you're still lost, you have no options. You must follow the four Cs technique: That is: Confess your problem to any ground station in the vicinity and don't try to sweat it out; Communicate clearly (don't panic!) and attempt to give as much information as you can

FIG. 4.4

Highway bridge over a railroad serves to confirm position.

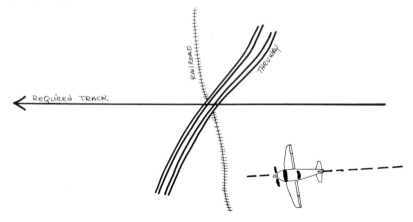

about where you think you are and what you can see below; Climb to a higher altitude where radio and radar contact will be improved; and finally Comply with what advice you are given. It may be galling to have to seek aid, but don't try to be a hero if you think you are out of your depth. And don't wait too long—if you haven't located yourself by the lost procedure outlined above within a reasonable time, start talking.

To conclude this brief look at navigation, make note of what you need in your navigation kit: First, two or three soft black lead pencils. Next, a navigator's ruler, which will be marked with statute and nautical mile scales, and scales for both sectional and World Aeronautical Chart (WAC) (1:1,000,000) charts. There's a good combination-type rule with a built-in protractor that some people like, but others prefer a separate protractor with which to measure off degrees. And of course you'll need that flight computer. Always have current charts, and make a point of having charts for those areas that are adjacent to the areas your charts cover for cross-country flying. Sometimes weather does change unexpectedly, and then it's nice to be able to take remedial action by flying around it. As you build up your experience, you'll find that radio charts—which give routes between the VORs—make an excellent addition to the flight kit of the VFR pilot. You'll find more details about them under the section on radios and instrument flying.

THE AIRMAN'S INFORMATION MANUAL

Before going on to the Federal Aviation of Regulations, let's take a quick look at the Airman's Information Manual (AIM). This is the operational flight manual, which contains all the information you need if you want to fly within the National Airspace System. You are expected to know what it contains.

The AIM comes in four parts. Part I deals with air traffic control procedures as they affect pilots and with fundamentals of flight operation. It contains educational and training information, and is titled *Basic Flight Manual: ATC Quarterly Procedures*. As the title indicates, it is revised on a quarterly basis.

Part 2 is an airport directory, covering all airfields, heliports, and seaplane bases in the continental United States, Puerto Rico, and the Virgin Islands, and is available for civilian use. (Curiously, although listing all facilities and services, it does not have details of communications—you have to buy Part 3 for that, which also contains a number of other items. That's because frequencies change more often than airports.) Part 2 also includes a list of Flight Service Stations and Weather Bureau telephone numbers, a listing of commercial radio stations that broadcast at 100 watts or more—which can be used for radio direction finding (ADF)—plus details of the latest Entry and Departure procedures for airplanes traveling abroad. Part 2 is issued twice a year.

Parts 3 and 3A contain operational data and Notams. In addition to including those airports with communications facilities, Part 3 also lists radio navigational aids and their frequencies, the preferred routes for instrument flights, and Standard Instrument Departures (SIDs) and Standard Terminal Arrival Routes (STARs), both of which are used by instrument pilots. Also listed are substitute route structure details, sectional chart bulletins, and special general and area notices. Finally, there are details of all new and permanently closed airports, with an increasing number of Area Navigation Routes (these will be dealt with elsewhere in this book). Part 3 is updated every four weeks; Part 3A is issued every two weeks.

Part 3A contains the Notams considered essential to the safety of flight plus supplemental information to Parts 3 and 4.

Part 4 contains graphic notices and supplemental data. It includes a list of abbreviations used in the AIM, details, of all parachute jump areas with the times when they can be used, VOR receiver checkpoints (both ground and airborne checkpoints are listed), and various special notices. Your instructor will show you the AIM, though you may decide that for your own private use, a commercially produced airport directory could be more helpful. Every FSS is supposed to have an up-to-date copy of the AIM; you should check Notams every time you fly. Part 4 is issued twice a year. Your cheapest recourse, however, is to subscribe to "J-Aid" service, produced by Jeppesen and Company.

FEDERAL AVIATION REGULATIONS

If the AIM is the publication that tells you all you have to know about being operational, the Federal Aviation Regulations (FARs) spell out the rules under which you may or may not operate. It is well worth reading in its entirety, since it will make you familiar with the wide world of aviation. You should know and have a good understanding of Parts 1, 61, 91, 93, and 430.

Part 1 is a list of definitions of the terms used in the FARs so that there will be no misunderstandings. If you don't understand what a particular term means, you can look it up in Part 1. Also listed are the abbreviations and symbols used. Part 1 also contains the rules of construction, but most pilots are not usually interested in these. They are the source of any number of tricky questions on exams, however.

Part 61 deals with the certification of pilots. It covers all types of certificates that may be issued, what skills are required to obtain them, their applicability, and what limitations are included. It also provides details of medical exams, and the keeping of logbooks. (Part 67 outlines the actual medical standards needed.) Everything is spelled out, whether you want to fly a balloon, soar in a glider, or pilot a helicopter. And it has just been revised, so make sure you're looking at a new copy.

Part 91 contains the rules for flight within the United States. There are changes from time to time. Perhaps the most important rule is 91.3 (a): *The pilot in command of an aircraft is directly responsible for, and is the final authority as to, the operation of that aircraft.* In other words, the captain of the airplane has the final say in flying—if he should need to break the rules in order to maintain the *safe* operation of his aircraft, the rules permit him this latitude . . . provided he can provide chapter and verse to support his decision. Flight rules are also covered, with details concerning both visual and instrument flight rules. Requirements for maintenance, how it is to be carried out, and by whom are also laid down in this section, as well as regulations concerned with the operations of large, turbine-powered multiengine equipment.

Part 93 provides details of special air traffic rules. The only portion of interest to those of us in the contiguous United States is that dealing with special VFR minimums for operation within a control zone. Parts 93.111 and 93.113 give these details and the control zones in which special VFR is not authorized. Subpart K of Part 93, High Density Traffic Airports, details special procedures to be followed at those airports that tend to have many commercial air carriers.

Part 430 deals with reporting accidents and what constitutes one. You'd be amazed at what doesn't. You can—if you are so minded— do a considerable amount of what you and I would call damage to an airplane without it being a reportable accident. You'll find the details in Part 430.2, which deals with definitions.

THE AIRPLANE ENGINE

The Piston Engine

Almost everybody knows that an engine is something you find under the hood of a car, but fewer know that the majority of car engines are piston engines, and fewer still know how they actually work.

Let's look at some examples. If the vehicle has two cylinders, there will be less vibration when the cylinders are opposed to each other; more precisely, when they are opposed in a horizontal rather than vertical plane. The smoothest twin-cylinder motorcycle in the world (BMW) still uses this method. If you use four cylinders, the best system—judging by the numbers around—is a horizontally opposed four, as used by Volkswagen and Porsche. Then you find V-6, V-8, and in the Jaguar and other exotic equipment, V-12, V-16, and even V-24 engines. But what you gain in smoothness and access to power by having a lot of cylinders, you generally lose in real power, or torque. Thus, some device—a gearbox in a car—is required to transfer the energy. To stay with motorcycles for a moment more, a big single-cylinder engine bike may not win drag races but will provide a reliable source of energy through many a long day without too much gear shifting. A multicylinder engine will provide rapid acceleration and demand a lot of work from its rider.

For an aircraft power needs are rather different. First, because we can fly like the proverbial crow, we don't have to reckon with going up and down hills and around corners, thus varying our speed. We need extra power to get us into the skies, and we need sufficient power to get us along at a reasonable rate once we've reached cruise altitude. Although this allows us to eliminate the gearbox, it does mean that we have to set a somewhat higher standard for our engine. It means that we collect a lot of energy with full throttle, which we use to take off, and that we throttle back just a little when we've reached cruise altitude. In other words, an aircraft engine tends to run at almost full throttle almost all its life.

This isn't all bad. An airplane engine doesn't have to deal with the problems of continually heating up and cooling down as a car engine does, so it doesn't need the complicated cooling system that most cars need. We can cool our airplane engine simply by passing air over it, as we cool most motorcycle (and Volkswagen) engines.

101

On the other hand, we must make sure the engine is properly cooled, or else it will heat up and stop. So in some aircraft we use special ducts to lead cool air onto the engine's surfaces when it is working rather hard—as on takeoff.

Because gasoline and air are being mixed to provide a combustible mixture, something is needed to do the mixing. Several methods are used; the two most common are the carburetor and the fuel injector.

There is an ideal ratio—approximately fifteen to one by weight— of air and gasoline respectively for an internal combustion engine. If you use less air, you get what is called a rich mixture, the traces of which may be seen in exhaust fumes. This wastes fuel (you consume more gasoline for the same return) and can hurt your engine by building up carbon deposits within the combustion chamber and by running your engine too cool. If you use more air, although you may be saving on fuel, you may be damaging the engine more seriously, as each expansion is accompanied by more heat than the engine may be designed to handle. Furthermore, if you use too much air, you get uneven explosions; no matter how ideally balanced your engine may be, you will quickly damage it.

The carburetor is designed to help solve this problem—with your help. You will find a simple lever to control the power and another lever—with a red knob—to control the ratio of air and gasoline. At sea level the mixture lever is pushed to fully rich and is leaned out (less rich) as the airplane gets higher. (Remember that the atmosphere gets thinner as you ascend. The higher you go, the less oxygen there is available for combustion purposes. Consequently, you use less fuel. Extended cruise figures are therefore usually quoted for the airplane's optimum altitude, frequently 10,000 feet.)

In a basic carburetor a fuel feed pipe leads the gasoline via a filter to a simple needle valve. This controls the actual amount of gas to be mixed with the air. The needle is also used to atomize the gas. Some carburetors use several needles that are fixed rather than moving, which are called jets. There are slow-running or idle jets, which keep the engine ticking over, main jets, for normal operation; and power (or boost) jets, when we need all the energy the engine can produce. Finally, there is the mixture control, usually an adjustable valve placed between the needle (or jets) and the combustion chamber, which regulates the amount of mixture entering the combustion chamber. The mixture of gasoline and air is then fed into a manifold, which provides a reservoir on which the various cylinders draw. Some very fancy designs of manifold exist for racing cars that, using various known behavior patterns of gases, increase the efficiency of the engine. The mixture is now ready to enter the cylinder (combustion chamber) proper.

Once a batch of air has been heated up and its energy used, it must be expelled. Entry and exit doors are thus needed, with both doors shut when the mixture is exploded so that its energy can be captured. The exit/entry problem is solved by using little doors called valves. A simple one-cylinder engine has an inlet valve to admit the mixture and an outlet valve, or exhaust valve, to let the mixture escape. The moving wall of the combustion chamber—the piston—is coupled so that it can return to squeeze the mixture. This is a slight simplification of what actually takes place. The piston performs a triple duty, first in expelling the air and second in compressing the mixture (compressing the gas/air mixture beforehand will produce more energy for the rotating crankshaft), and finally transmitting energy to the crankshaft.

When you turn the key to START, the engine is turned over by a starter motor powered by the battery. The mixture is then drawn into the cylinders by the opening of the inlet valves and the movement of the piston away from them. Meanwhile, a spark to ignite the mixture is prepared by the magneto, which is set to dispatch its electrical energy to the spark plug at an appropriate moment.

Returning for a moment to the compression of the gas/air mixture, it is in this connection that such terms as compression ratio and octane rating arise. Compression ratio has to do with the volume of the mixture as it enters the combustion chamber and the degree to which it is compressed by the piston. An 8:1 compression ratio means that eight units of mixture are compressed to one unit. Octane rating has to do with the property of gasoline. The oil removed from the earth needs refining before it can be used. Lower compression ratios require a less-refined gasoline (low octane) that is cheaper but provides less power. Higher compression ratios, which provide more power, require more-refined gasoline (high octane), which in turn costs more.

The point is that engines—especially airplane engines—are designed to use a specific type of gasoline. If your engine is rated for 91-octane gas and you use 80-octane gas, you will get uneven burning, which will produce detonation (a very quick explosion). On the other hand, if you use too high an octane gas continuously, you could, as mentioned earlier, cause engine damage from excessive heat. If you must choose between too high and too low octane, choose the higher. Excessive heat is less damaging than violent uneven explosions.

In a car the spark comes from the coil; in airplane engines it comes from a similar sort of device called the magneto, which was mentioned earlier. Both an ignition coil and a magneto build up a high-tension current that causes a spark to jump across the spark plug, igniting the compressed mixture in the cylinder.

The mixture explodes, pushing the movable wall (the piston) away from the spark and transferring its moving energy to the crank shaft. As the explosion dies down, the exhaust valve opens and the spent gases are forced out as the piston moves back toward the spark plug. If you've been counting the number of moves, you'll have counted four, which is why certain types of engines are called four-stroke engines. (The first stroke was a downward movement of the piston, which assisted in drawing in the mixture; the second stroke was an upward movement, which compressed the mixture; the third stroke was the downward movement caused by the explosion; and the final stroke was the one that expelled the used gases.)

We said that the energy from the exploding mixture was relayed by the piston to a shaft. This is not as easy as it sounds, since there

FIG. 4.5

Inlet valve (a1) is where the mixture of gas/air enters the cylinder. Exhaust valve (a2) takes the burnt mixture to the exhaust pipe. In the interests of safety each cylinder has two spark plugs (k1 and k2), each independent of the other. Crown of piston (b) receives the force of the sealing so that the explosion doesn't get beyond the piston. Skirt of piston: (d). Gudgeon pin (e) secures piston to connecting rod. Small end of the connecting rod (f) is provided with its own bearings to reduce friction. Big end bearing: (h). Journal (j) provides a miniflywheel effect in addition to providing for the transfer of energy from the combustion chamber to the crankshaft (i). In light aircraft the crankshaft is attached to the propeller at one end and to the flywheel at the other.

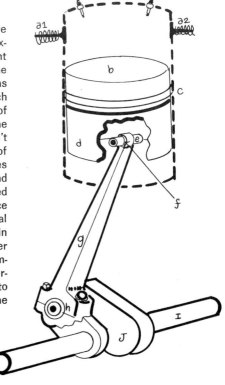

104

is limited space in the engine. The shaft, or crankshaft, is arranged in such a way that the downward movement of the piston is converted into a rotary motion. (The piston is attached via a connecting rod to a journal, a part of the crankshaft that is displaced away from its center. The reason for the displacement is obvious—to collect the energy and permit the piston to provide the rotary motion through the big end of the connecting rod. (See Figure 4.5.)

The valves are controlled by springs that act to keep them closed. They are opened at the appropriate moment in the cycle by a lever, called a rocker arm, which in turn is activated by a push rod, whose movement is controlled from a camshaft. The camshaft is fitted with raised portions called cams, which make the pushrod do its work. The cams are eccentric in shape because the valves need to be open only a certain amount of the time. The exhaust valve is open only when we want to exhaust the cylinder, and the inlet valve is open only when we want to introduce mixture into the cylinder. (See Figure 4.6.)

The spark is needed only some of the time during the cycle, and this is controlled by a distributor, which sends the high-tension current to the various spark plugs as it is needed.

FIG. 4.6

At (a) the gas/air mixture enters. At (b) the inlet valve is now shut and the piston drives upward, compressing the mixture for a better explosion. At (c) the spark plugs ignite the compressed mixture. At (d) the residual energy from the explosion brings the piston up, driving out the exhausted mixture.

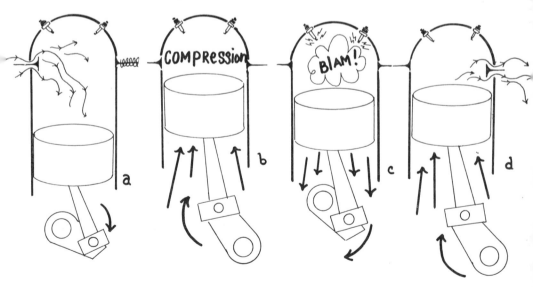

Because there is a lot of friction, the various parts need to be lubricated. Two systems are employed: The first, which simply depends on oil being thrashed around the moving parts, is called splash lubrication; in the second, special oil lines pressure feed the lubricant to the various points where it is needed. The oil is filtered each time it goes through the system to remove any particles of metal or carbon it may have picked up on its journey.

Finally, there is the problem of cooling the engine. Internal combustion engines need to be cooled—of the energy produced with each explosion, only one-third is converted into usable power. The remainder is heat. Although liquid cooling has been used successfully on airplane engines, it is cheaper and simpler to use air cooling.

The Turbine Engine

The earliest uses of jet propulsion were probably the rockets used by the ancient Chinese. And of course the Greeks knew about it too. Jet propulsion can be demonstrated very simply by taking a toy balloon, blowing it up and then releasing it. The balloon flies away. The reason the balloon moves away from the escaping air is that it is reacting to it, not because the air is hitting the atmosphere and forcing the balloon along. The balloon moves in direct reaction to the jet.

Jet engines operate on the same principle, only they do it by taking air and fuel, compressing it, igniting it, and then expelling the gases. The big difference from piston engines is that all this is taking place simultaneously within the jet engine in its different parts. This results in the distinctive smoothness of operation common to gas turbines, because there is no up-and-down, in-and-out movement involved. (See Figures 4.7 and 4.8.)

The present type of gas turbine, which utilizes axial flow rather than centrifugal flow, is the result of design work undertaken by Dr. A. A. Griffith at the Royal Aircraft Establishment, Farnborough, England, as early as 1926. Two years later another Englishman, Frank Whittle, proposed the use of the gas turbine for aircraft, and his patent was granted in 1930. Similar work was being undertaken by Dr. Pabst von Ohain in Germany during the 1930s, and the first airplane to fly with a turbojet engine was the Heinkel He 178, just days before the outbreak of World War II in Europe.

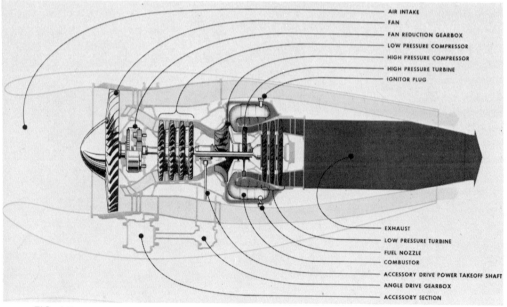

AIR INTAKE
FAN
FAN REDUCTION GEARBOX
LOW PRESSURE COMPRESSOR
HIGH PRESSURE COMPRESSOR
HIGH PRESSURE TURBINE
IGNITOR PLUG

EXHAUST
LOW PRESSURE TURBINE
FUEL NOZZLE
COMBUSTOR
ACCESSORY DRIVE POWER TAKEOFF SHAFT
ANGLE DRIVE GEARBOX
ACCESSORY SECTION

FIG. 4.7

The Garrett/Airesearch TFE 731 turbofan is a relatively high bypass front fan engine. Air enters the engine at left, most of it going through without anything happening to it (hence bypass). The remainder goes into the four-stage axial compressor (which heats it), then on to the centrifugal compressor on its way to the combustor, where it picks up the additional heat energy of the burning fuel/air mix.

FIG. 4.8

Cross section of the Airesearch TFE 731-2 turbofan engine

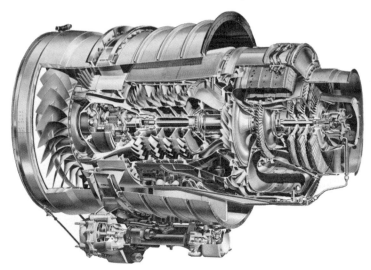

The centrifugal compressor-type turbine was developed from piston-engine superchargers, devices that force the mixture into the cylinders at a pressure higher than atmosphere pressure. In the early centrifugal-type turbines the mixture was fed into the center of an impeller that revolved at high speed, forcing the air outward with a rapid raise of pressure. This high-pressure air was sometimes fed to a second impeller, where the process was repeated. The reason for compressing the air—which by this time had reached a temperature of more than 200° C. by reason of compression alone— was in the interests of fuel economy.

It was the need to have higher compression ratios that led to the return to the axial flow compressor, in which a gradual increase of pressure is obtained by using a series of rotating airfoil-shaped discs. These discs are nearly works of art and have a very complicated section that must not only provide gradual compression but must also withstand and be able to consume any birds that might be ingested. Gas turbines undergo extensive testing, and birds are shot into the engine inlet at high speeds. An engine such as the Garrett Airesearch TFE 731-2 Turbofan can ingest a four-pound bird fired into the engine at nearly 200 mph and keep running.

The highly compressed air is now directed to the combustion chamber, usually called the flame tube, where a number of fuel sprays introduce finely atomized kerosene into the air. This is ignited by a powerful spark, and once ignition is achieved the spark is switched off, as the engine will keep burning (unless there's a flameout due to such things as ice, or foreign object ingestion or compressor stall, which happens from time to time).

Modern gas turbines make use of several sophisticated techniques in order to provide greater economy and quieter operation. These include by-passing some of the compressed air around the combustion area in order to lower the heat of the gases at the turbine. Temperatures in excess of 2,000° C. are recorded in the hot part of the engine. The turbine absorbs a considerable amount of energy in propjet aircraft, which is used to drive propellers as well as the various parts of the engine. In pure jet engines most of the thrust energy of the expanding gases is used as a high-speed exhaust from the rear end of the engine though even more thrust comes from the by-pass of a large fan-jet engine, producing a forward reaction imparting motion to the airplane.

Although jet engines tend to cost more than piston engines, their astonishing record of reliability and the large number of hours they can operate between overhaul mitigates against the initial large

The Interceptor 400

expense. Several popular aircraft make use of the gas turbine, including a single-engine, four-seater monoplane called the Interceptor 400 (Figure 4.9). Faster than most twin-engined airplanes, the Interceptor 400 is pressurized (like an airliner) and cruises at 280 mph for a range of nearly 1,200 miles. It doesn't need long runways either—takeoff roll is less than 900 feet. But it costs about $145,000.

PART III

Cessna 150 cockpit: Mainstay of the rural FBOs flight training program, earlier 150s earn their keep all over the world. The instruments from top left to right are: airspeed indicator (ASI); directional gyro (DG); artificial horizon (A/H). Below left to right: turn-and-bank (shown is the needle/ball arrangement) and vertical speed indicator (VSI). A clock is placed below the turn-and-bank. The old-fashioned Narco radio provides a clue to the aircraft's age.

5

Careers and the Rating Game

The idea behind ratings is older than the medieval craft guilds; its intent is to assess competence. There are certain standards that pilots and the airplanes they fly must meet. Although standards of aircraft design could be improved—for example, there is fairly clear evidence that visibility from the cockpit of the DC-9 could use improvement—the overall standard of the nation's pilots is actually pretty good and is likely to get better.

In the recent past many pilots have been making it a practice to have an annual flight check, just to ensure proficiency. Under the new FAA rules that took effect November 1, 1974, no person may act as pilot-in-command unless, within the preceding twenty-four months, he has completed satisfactorily—and logged—a flight review given by a certified flight instructor. Alternatively, he may have passed a proficiency check for a pilot certificate, rating, or operating privilege given by the FAA. It is not actually a check flight, since there is no way a person can "fail." But it may not be logged unless it has been completed *successfully*.

As we have seen, the absolute minimum required to fly in the United States is a third-class medical when you're starting to fly, which doubles as a student's license. The only trouble with this is that you may not fly as pilot-in-command unless signed off by your instructor. Nor may you legally carry passengers, unless they are pilots, in which case you would designate one of them as pilot-in-command.

The student license is not of much use, then, unless you are actually learning to fly. After the student license the bare minimum is the private pilot certificate, which entitles you to fly under visual flight rules and to carry passengers. This certificate is endorsed for flying an airplane single-engine land (ASEL). If you want to fly a floatplane, amphibian, or twin, you have to demonstrate your competence in such a vehicle; then the appropriate notation will be added to your current certificate.

If you're confused, it may be helpful to examine first the types of certificate that may be issued to see what privileges they confer. You may not act as pilot-in-command (or, as the book says, "in any other capacity as a required pilot flight crew member") of a civil aircraft of U.S. registry unless you have in your possession a current pilot certificate issued to you (FAR 61.3 [a]). And even if you have such a certificate, it's no good unless you have an up-to-date medical certificate. (A note about medical certificates is to be found at the end of this chapter.)

Licenses—or certificates, as we should really call them—are divided into private, commercial, and air transport certificates. The air transport certificate is actually a rating that is added to the commercial certificate, but it is usually spoken of as if it were a certificate in its own right. The new commercial certificate, which came into force November 1, 1974, requires a total flight time of 250 hours, compared with the 200 hours required earlier. It also stipulates an instrument rating, which prohibits the carriage of passengers for hire on cross-country flights of more than fifty nautical miles or at night, as a requirement—with a dispensation available for agricultural pilots, for example, who may not want the instrument rating.

The commercial certificate also requires a minimum of ten hours in a "complex" airplane, which means an aircraft with retractable gear, a flap mechanism, and constant speed prop. The number of hours of flight instruction is also increased from twenty to fifty hours, though of the total of 250 hours, 50 may be logged in a ground trainer.

Still, if you acquired your commercial certificate before November 1, 1974, your certificate will always be good, there being a grandfather clause exempting existing commercial pilots from the instrument rating rule and the limitation. The commercial certificate, as you can see, requires more experience and skill than the private pilot certificate. The air transport rating (ATR) is actually tacked onto the commercial certificate and requires considerably more knowledge and experience than the ordinary commercial certificate.

Category ratings are further subdivisions of the private, commercial, and ATR certificates. These subdivisions are airplane, rotocraft, glider, and lighter-than-air aircraft. Within each category there are further subdivisions: Airplane class ratings may be divided into single-engine land, multiengine land, single-engine sea, multiengine sea, single-engine land and sea, and multiengine land and sea. Rotorcraft ratings are more simply divided into gyroplane and helicopter subdivisions. The difference is that in a gyroplane the main rotor is not powered during flight, although it may be powered for a jump takeoff. The subdivisions for lighter-than-air ratings are airship and free balloon.

Finally, the name of each type of large aircraft and each turbojet-powered airplane for which a pilot is rated must be placed on the certificate. Large aircraft is defined as an airplane with a gross weight in excess of 12,500 pounds. Helicopter type ratings are also issued for each type of helicopter in the case of ATRs. It will be seen that there are a number of possible certificate permutations, all of which, alas, spell money in terms of obtaining them.

For this reason it's a good idea to determine ahead of time what kind of flying you will be doing and for what reasons. If you merely enjoy flying for the music it pours upon the soul, as it were, your needs will possibly be satisfied with an ordinary private pilot certificate and some lessons in aerobatics. But if you are going to employ your skills in some other areas, perhaps for personal transport or'

FIG. 5.1

Schweizer's 2-33 sailplane is the successor to their popular 2-22 model in which many of America's top soaring pilots took their first flight.

even to earn a living, then you will need to think more closely about what certificates you should get and in which order you should go about getting them.

As an absolute minimum, each pilot should acquire a private certificate with an instrument rating. This offers fair versatility and at the same time permits the pilot to optimize the use of his airplane. Having got these two, he or she might like to think about whether gliding or seaplanes don't offer a chance of personal expansion in using the skies. Or perhaps you might take a course in aerobatics. The instrument rating is useful, not simply because it enables you to make use of the Air Traffic Control (ATC) system but because it also increases the experience of the happy-go-lucky VFR-only pilot. The one complements the other and tends to produce a pilot with very real command ability.

Quite apart from increasing the utility of your machine—according to some theorists, from around 55 percent to more than 90 percent—the instrument rating enables you to become more proficient, that is, if you are prepared to maintain the additional skills you are obliged to learn to obtain it in the first place. For the problems of flight remain the same: You must keep the airplane the right way up, you must keep it going where you want it to go, and you must keep track of where you are all the time. The only real difference between VFR and IFR is that with the latter, you acquire the information you need to perform these tasks from your instruments. The good IFR pilot has the ability to think creatively about this information, so that he always has a clear mental picture of his position.

Although nearly one in five pilots is instrument rated, only a fraction of nonprofessional pilots fly by instruments on any sort of a regular basis, although the number is increasing. One reason for this is that teaching methods are getting better all the time. Flight simulators at even quite modest schools have built-in turbulence, for example, and can also copy the ATC environment in which instrument flight takes place. Furthermore, with a simulator it is possible to "stop" the airplane in midflight, climb out of the cockpit, and actually see—on a tracing board just outside the cockpit—how a mistake was made during a particular maneuver. This is especially helpful before actually getting into the air.

Perhaps the most difficult problem for the would-be instrument pilot is handling the ATC environment. Various new procedures have to be learned, such as copying clearances correctly, both on the ground and in flight; entering a holding pattern from any direction; or finding the Instrument Landing System (ILS) after Control has vectored you by radar to your final approach, but has managed to place you smack on top of the outer marker and two hundred feet too high!

THE INSTRUMENT RATING

As in the private pilot certification, the instrument rating requires a written examination and a flight test. A private pilot wishing to take the instrument rating must still have acquired 200 hours of flight experience before being eligible (there's no change in this part of the regulations). Until November 1, 1974, the pilot must have logged at least one cross-country trip of 200 nautical miles in simulated or actual IFR (250 nautical miles in future); following November 1, 1974, the pilot must be able to demonstrate VOR, ILS, ADF approaches, although the ILS and ADF approaches may be demonstrated in a simulator.

A minimum of forty hours instrument time is required, of which at least twenty must have been logged in an airplane. Of these twenty hours, at least fifteen must have been instrument flight instruction given by a flight instructor with a certified instrument instructor rating. The remaining time may be taken in a simulator or an airplane. Slightly different rules apply for helicopter pilots in that at least five hours of the required fifteen hours must include instrument flight instruction in an ordinary airplane.

The written exam is quite difficult, consisting of multiple-choice questions carefully designed to trap those who do not know the subject. As in the private and commercial certificate written exam, one answer will usually be obviously wrong and a second less obviously wrong. Of the two remaining, both may appear correct, but a closer examination of these two will demonstrate a falacious quality about one, which leaves you with your answer.

The areas covered in the written exam include radio navigation systems and procedures, instrument landings systems and procedures, and radio procedures. You are expected to be familiar with the several means of landing in weather, including landings by means of nondirectional radio beacons (ADF), omnidirectional beacons (VOR), and so on. Additionally, you should have a better than nodding acquaintance with basic meteorology, including the characteristics of air masses and fronts, and how to work out a satisfactory alternate airport to land at if necessary.

The FAA's own *Instrument Flying Handbook* provides an extremely comprehensive starting point for the instrument rating, and if you supplement this with Robert N. Buck's *Weather Flying* and Richard L. Taylor's *Instrument Flying*, you'll be started in the right direction. *Flying* magazine's *Guide for Instrument Flying* also provides some useful ideas. Jeppesen also offers a good home study course. Several other good source materials exist for the instrument rating, a number of which are listed in the bibliography. Shop around and find what you think suits you best. Instrument flight is discussed more fully in Chapter 10.

MULTI-ENGINE RATING

Adding an extra engine to your airplane is like having money in the bank, according to those who value the theoretical safety of an additional fan. Unfortunately, it also costs more. Twice as much gas, twice as much engine and propeller maintenance, and usually higher insurance premiums are required, as the average twin-engine aircraft costs quite a bit more than the average single-engine aircraft.

There's no doubt that there are numerous satisfactions in owning a twin-engine aircraft, but the additional cost should be an important factor in making a decision. Also some twin-engined aircraft possess certain characteristics that make them less desirable. The usual time these problems become apparent is when one of the engines fails. The National Transportation Safety Board (NTSB) in 1973 published an interesting study of accidents involving engine failure/malfunction for the period 1965–69. This study showed that twin-engine aircraft had half as many engine-failure accidents as single-engine aircraft—2.3 accidents per 100,000 hours, as compared with 4.6 per 100,000 hours for singles. But with twin-engine aircraft four times as many engine-failure accidents were fatal than with single-engine aircraft.

Although the statistics talk of engine failure, fuel mismanagement was a probable cause of engine failure in nearly 40 percent of these. Interestingly enough, the aircraft that stood out best was Cessna's 172. Its rate of serious accident after an engine failure was one in every 800,000 hours of flight, better than many twin-engine aircraft.

The reason that aircraft originally used more than one engine was that early engines were rather limited in power. And even today some twin-engined aircraft at gross weight won't climb much on only one engine even near the ground, while others with one engine out at high altitudes have to fly at much lower altitudes compared to their lofty "twin" performance peaks. There's always the question of the remaining operating engine's thrust and the airplane's resultant behaviour at or near stalling speeds. Fortunately, design engineers are beginning to solve these problems, bringing us centerline thrust (from Cessna in the Skymaster 337 series) or C/R (contrarotating) propellers from the 'Piper stable. (See Figure 5.2.) Centerline thrust solves the problem of a failed engine by mounting the two engines in tandem so that no problem with asymmetrical thrust can arise. The second solution—making the propellers contrarotate relative to each other—also solves the problem of asymmetrical thrust, which occurs when both engines rotate their props in the same direction. (See Figure 5.3.) The point that should be understood about twin-engine aircraft is that they do require extra proficiency in their handling; and, if they are to provide the extra safety that theory suggests, recurrent training in handling them with one engine feathered is vital.

Although there's no minimum number of hours of experience required for the multi-engine rating, examiners like to see at least 10 hours logged. Because there's also no written test, examiners usually conduct a fairly extensive oral quiz in addition to the flight test, which in turn emphasizes the techniques required for flying with one engine out.

Once you get the hang of it, flying with only one engine operating is not difficult. What you have to do is to be able to recognize immediately which engine is not working (the air pressure on the rudder and visual inspection tells you this) and provide opposite rudder trim. Then raise flaps and retract landing gear, if out, to clean up the airplane in order to get some altitude.

If you're intending to fly multi-engine aircraft all the time, it is good operating practice to check yourself out on your single-engine

FIG. 5.2

Piper's popular twin Comanche, now fitted with contra-rotating propellers.

procedures fairly frequently so that should an emergency ever occur, you'll be able to do it by the numbers. If you intend to fly multi-engine and instrument, a separate flight check in the multi-engine aircraft is required for your instrument rating.

SEAPLANE RATING

As with the multi-engine rating, no minimum number of hours are required for the seaplane rating, and there's no written exam. Still, you can expect to take up to ten hours learning how to maneuver an airplane on the water, taking off, and landing. If you have a floatplane, every small lake becomes a landing strip, and if you happen to have the amphibious type of float (with wheels that retract into the floats) you can also land and takeoff from regular airstrips. (See Figure 5.4.) The same applies to flying boats proper.

You can expect a 20-mph penalty on cruise speed with floats on an ordinary airplane, plus some reduction in useful load. Against that you can add the considerable extra security in wilderness areas. A seaplane rating opens up a whole host of areas denied to others, fliers and nonfliers alike.

FIG. 5.3

Cessna Skymaster 337

FIG. 5.4

The seaplane rating gives you access to wilderness areas, as shown by this Cessna airplane and crew. The lightweight Rushton canoe straps easily to one of the floats, providing local water transport—and for fishing, too.

THE COMMERCIAL CERTIFICATE

This is the rating you have to have if you want to fly people and get paid for doing it. Changes in FAA requirements are in effect as of November 1, 1974, and the biggest change between the old regulations and the new ones is that an instrument rating becomes mandatory. For Ag (agriculture) pilots and others who don't want to bother with an instrument rating, a special limited certificate is available that prohibits the carriage of passengers for hire on flights of more than fifty nautical miles or at night.

Although most pilots are not too interested in obtaining the commercial certificate for purely monetary reasons, many pilots look to the commercial certificate as setting a standard in skill, as the commercial pilot must be able to do everything that the private pilot can do, only better. There are also new areas in which this skill must be demonstrated, including, until November 1, 1974, some instrument proficiency and precision flying, such as maneuvering at minimum controllable airspeed, stalls from all normally anticipated flight attitudes with and without power, right and left 720-degree power turns, and such semi-aerobatic maneuvers as chandelles, lazy eights, shallow and steep pylon-eights, and gliding spirals around a point. In addition, three accuracy landings within 200 feet of a designated mark are required, which may be included in the several types of takeoffs and landings that form a part of the flight test.

It all sounds formidable, but it isn't as bad as it seems. For one thing, before you can apply you must be at least eighteen-years-old (compared with sixteen for the private certificate) and must have at least 200 hours of flight time. (The new rules will require 250 hours.) There comes a point where the commercial certificate begins to be a challenge to your skills. If the actual flying is not as difficult as it seems, some of the book learning can be confusing if it is not approached in the right way. Because much of it covers the same ground as the instrument written exam, there's a good case for doing the two at the same time.

One final note—to use your commercial certificate (indeed, to take the flight test for it) you'll need to pass the second-class medical. This is stiffer than the medical for a private pilot certificate, and more emphasis is placed on being able to see properly.

Those interested in the possibility of flying for the commercial companies or corporate aviation will need at least a minimum of a Commercial Pilot Certificate with instrument rating, plus the written test of the ATR and evidence of several hundreds of hours flown.

Airline management in this country has very largely followed a policy of staffing their flight rosters with pilots trained by the military. If, as a civilian pilot, you're considering working towards the right seat of a commercial jet, think carefully. You may be better off in other fields of flying, all rewarding in their way if lacking the patina of an airline position.

A second problem you face—and this extends to corporate flying —is the burgeoning of employment agencies for pilots and aircrew. These agencies call themselves "personnel firms" and usually employ free-lance industrial psychologists who all too frequently don't know an aileron from a tailwheel, but whose position on the "team" ensures an "expert" evaluation to the finding and placing of a flight crew. The point is that military pilots—unless discharged unfavorably—can almost always be guaranteed a job, since the military has already subjected them to the best psychological testing the country can buy—before they ever began to learn to fly military style.

While some of these agencies do fill an important need for smaller companies who cannot afford a professional employment (or personnel) department, the whole business of finding and hiring personnel is mainly to find reasonable talent which while not brilliant will not make waves. It's the Peter Principle in operation.

Most psychological testing today is still very crude and is aimed more closely at finding whether a person is willing to not rock the boat. Psychologists have, for the most part, a very special interest in the boat not being rocked, since if rocked, they might be the first people to have to swim. Conformity to existing mores is their bag, and in terms of airline employment—and other employment—this means finding people willing to whore for the system.

And this means that regardless of the fancy language in which the evaluation report is dressed, what is really being talked about is how closely you can be said to be compatible with a particular social grouping. Careful reading of the questions will undoubtedly help you to pass some of the less obvious pitfalls on your way— though there are cleverly worked catchall ("Catch-22") type questions that you'll answer wrong whichever way you choose. You'll also be wrong if you don't answer. A reasonable fund of general knowledge also helps.

If you've been thinking about working up to this summit, you'll undoubtedly have heard of the "dreaded" Stanine test. This was evolved by psychologists from a test program designed during World War II and of which the British novelist, Compton Mackenzie, wrote an amusing book. A typical question in this test concerns the paintings in the Sistine Chapel. The question is designed to find out what you know, and tells you rather more about the writer of the test than he realizes.

1. Unless you were interested in art you would not normally know that the frescoes in that chapel were painted by Michelangelo and several other artists.
2. If you were to display your precocious fund of knowledge and point out that:
   (a) Sistine has to do or is about any of the popes named Sixtus;
   (b) it is more commonly associated with the famous edition of the Vulgate (Latin version of the Scriptures) and originally issued by Pope Sixtus V in 1590—some twenty-six years after Michelangelo's death —and ultimately revised a couple of years later by Pope Clement VIII;
   (c) the chapel was actually built by Pope Sixtus (who died in 1484); or
   (d) Sistine is also the name of a color, an attractive greenish-blue of medium brillance and low saturation that enjoyed a certain vogue among painters of the Flemish school, the founders of which were Hubert and Jan van Eyck . . .

Well, the chances are that you might flunk. As an airline pilot you're just not supposed to be a repository of general knowledge, and if you are, everyone else is going to feel a lot happier if you appear somewhat dumb. Captains of industry do not like to feel like ignoramuses, since the only thing they know is what makes Sammy run.

As for knowing all that stuff about popes—well, it's decidedly unecumenical.

One of the more on-the-ball companies that still uses the Stanine test has modified this question to make it slightly more relevant to aviation, asking instead: what terminals form the three points of the Golden Triangle? (Answer: Boston, New York, and Washington, D.C.). Yet it is still difficult to see the relevance of a journalistic cliché in aptitude testing for a job in the front of the airplane. To whom, after all, is the triangle "golden?" Not to the pilots who have to fly the route, and judging by congressional feelings, not to a number of passengers. Certainly, not to railroads. Possibly then only to airline boards and stockholders—do you see now how much you must surrender of yourself? Golden triangle? Hardly!

The Stanine test also includes a number of quizzes of a type employed in Britain some years ago to evaluate eleven-year-old children to see whether these would be entered into an academic stream, or modern or technological course at high school. These tests measure your ability to complete sentences (to ensure you are not *that* dumb): *If cat is to mouse, then rat is to . . . ?* Multiple choice: *The leaves on the yuletide tree are red/white/blue/green/ gold/other color. . . .* And what are pompously called "problems in spatial relations." This is jargon for being able to identify a commonality within shapes, from simple triangles and circles to polygons and polyhedrals, colored black and white and which must be paired. If it's any consolation, it seems that people's IQs improve with age.

If you are looking for a job with an airline you must realize that you have got to be "all things to all people." You have to be singularly stupid to be the head of any organization, so involved in your own ego trip that you cannot see the wood for the trees. If you are using the organization to provide you with the necessary money to live your own life then you will know what compromises are required. If you find it all faintly ridiculous to conform to these primitive and tribal imperatives in the latter part of the 20th-century, or if it leaves you feeling moderately debased as a person, then you will undoubtedly be happier in some other field of flying. Don't worry about what the shrinks say about you, your degree of introspection, your ability to adjust to circumstance, and how much you need to dominate and be dominated. This is very moot.

Solon had it in a nutshell—Know Thyself, he suggested. You do not need to feel you have been put down by your refusal to play t-h-e-i-r game. Generally, the civilian pilot looking for an airline or corporate position should meet the following requirements.

Sex: Male (Unfortunately females may almost just as well forget it).

Age: 21 to 30 preferred, although exceptionally qualified pilots with extensive flight experience may be considered up to age 32.

Height: 5′ 9″ to 6′ 4″ usually, although shorter pilots have sometimes made it.

Vision: 20/20 uncorrected with no color deficiency.

Education: At least two years at an accredited university or college.

Medical: First-class without limitation; FCC radio-telephone permit (restricted).

Most pilots being hired today by the airlines have higher educational qualifications than those listed and usually more than 1,000

FIG. 5.5

To speed pilot training, airlines use sophisticated simulators capable of duplicating all required aircraft handling characteristics.

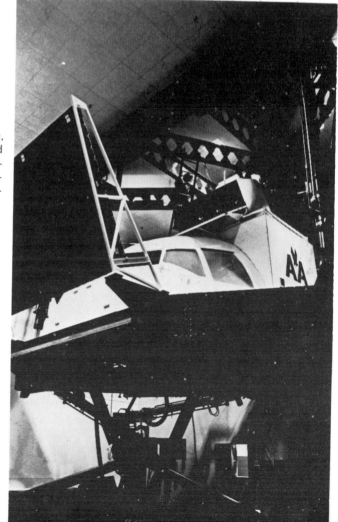

hours flight time. Pilots with 500 hours in jet equipment or 500 hours in large aircraft (over 12,500 pounds, multiengine) may sometimes be considered.

The energy crisis provided a much needed excuse to airline companies to cut back their schedules, which because of duplication were increasingly less profitable. Still, it would be dangerous to predict the actual decline of airlines though certainly their growth rate has been dramatically slashed. One reason was the ill-considered decision to go ahead with the use of jumbo jets. And if the airlines are in trouble today it is because they have too many seats with too few passengers. In order to fill those seats many airplanes will have to be moved into charter service. There is also the burgeoning shortage of jet fuel, the increasing discomfort of long waits for baggage and searches for weapons and bombs before boarding.

From a woman's point of view, the chances of an airline job are still rather slim. The only two American women pilots—both highly qualified—owe their jobs to themselves and to companies sufficiently oriented toward the public-relations aspect of having a token woman pilot on the staff. Almost all the airlines are intensely male chauvinist. By and large, if a woman wants to make a career with a commercial airline, she may have to buy her own company or establish a new one—certainly in the United States. She will also find that her talents will be mainly directed toward teaching and that other opportunities will be closed off to her. This is not because women are somehow inherently worse pilots than men, although this prejudice is behind the reasons offered. Incidentally, there are very good women pilots flying with commercial airlines at the present time in both Sweden and England.

THE AIRLINE TRANSPORT CERTIFICATE

The airline transport certificate is the Ph.D. of aviation; to get it you must be at least twenty-three years old and very good indeed. There is a complicated written examination, and before you can take the flight test, you must show that you've logged at least 1,500 hours of flight time within the past eight years, with a minimum of 250 hours as pilot-in-command. You must also have a first-class medical. In addition, you'll need 75 hours of instrument time, of which 50 hours must have been in actual flight, 100 hours of night flying, and at least 500 hours of cross-country flying. The first-class medical, by the way, is a difficult one to pass.

FIG. 5.6

Modern jet cockpit is really only a scaled-up version of a light trainer.

CORPORATE AVIATION

In terms of salaries and added bonuses, the corporate field is as good, or almost as good, a field of employment as the airlines. A captain of a corporate Boeing 727, for example, could expect to earn anywhere from a minimum of $25,000 to $42,500 or more. DC-9 pilots make a bit less, and Gulfstream II pilots make almost as much. Pilots of less exotic equipment, such as the four-engine Lockheed Jetstar, make almost as much, although Sabreliner captains and copilots apparently get left out in the cold with a minimum of $12,000 plus to a maximum of $31,000, according to a report in *Professional Pilot* magazine. On the other hand, Learjet pilots can make quite a bit of money, up to nearly $40,000 for captains, with rather less for copilots (a maximum of $20,000, if they're lucky).

In the turboprop field the best money goes to captains (and copilots) of the Gulfstream I (up to $35,000 for captains and

FIG. 5.7

Rockwell Sabre 60

$20,000 plus for copilots). Drivers of the Twin Otter take the bottom rank of salary. Conventional twin-engine aircraft are divided, curiously enough, between the Convair series and the Aero Commander, the Convair having a slight edge. With light twin-engine aircraft, lowest salaries go to pilots of Apaches and Senecas, who also make the top salaries in some circumstances, at around $20,-000. Pilots of two heavy single-engine aircraft, the Beech Bonanza and the Helio Short Takeoff and Landing (STOL) Courier, make from $9,000 to $14,500.

In the charter field rates are slightly down and are well down in the commuter, or third-level airline, field. Even helicopter pilots don't do too well here, Boeing Vertol pilots' salaries being quoted at from $10,000 to $13,500 and copilots' salaries a meagre $7,000 to $8,500. A corporate pilot in an Alouette does rather better, with a maximum of just under $20,000; a Jet Ranger captain can make a salary in the mid-$20,000 range.

Still, paycheck figures don't tell the whole story. More and more corporations are being pushed into aviation, not merely to stay ahead of their competitors but simply to keep up with them. And the company that decides to own or rent its own airplane(s) is in a much better position to improve the cost effectiveness of its executives, of which one of the most important factors is flight scheduling.

The commercial companies almost never have an aircraft heading in the right direction at the right time for the busy executive.

128

With a corporate airplane an executive can finish his meetings and then be whisked on to the next one—direct. No business about changing aircraft and waiting for baggage for him. The versatility and practicality of small jets and turboprops, such as the Japanese Mitsubishi, can also be seen in the type of advertising the Cessna company has been using for their little Citation. Nor is it totally without interest that some 80 percent of *Fortune* magazine's top 500 corporations either own or periodically lease one or more aircraft.

For the corporate pilot flying is rather different from the type of flying undertaken by the airline pilot. The principal difference lies in the irregularity of destination and time, as the corporation may want its aircraft to fly anywhere and at quite short notice. Conditions of employment are by and large good, but they do vary from company to company and depend very much on management's own utilization of the aircraft. However, you may expect the same high professional standards in the corporate field—including regular checkrides—that you'd find with the airlines.

The types of aircraft used in corporate aviation also vary, and you will today find increasing use of light twin-engine aircraft, such as the Cessna 310 or 401, the Piper Aztec, and the Beech Baron. In smaller companies it is not unusual for the company airplane to act as an air taxi, taking executives to a relatively nearby main airfield where they will board a commercial airliner to complete their journey.

Perhaps the most irritating factor of corporate piloting is its irregularity. At first sight this may seem strange, as we normally think of change as being good. However, it should be realized that the corporate pilot is frequently his own navigator, flight engineer, ground maintenance officer, plus taking on any other job when required, as far as his aircraft is concerned.

Another problem connected with scheduling according to company needs is aircraft maintenance. Many times an aircraft has been put in for its 100-hour check or annual when it is required by a corporation executive for an immediate flight on very important business. The solution to this problem is a progressive maintenance program, but outside of the airlines few servicing companies have worked out such maintenance schedules that keep everyone happy. Thus the pilot himself may have to work out such a system in order to optimize the usefulness of the aircraft.

Fringe benefits can be good. Perhaps the most appealing are the retirement programs, good vacations, and even options to buy company stock at favorable prices. Frequently the job of pilot may be only part-time, and the company pilot may find himself as an administrative assistant to a senior executive when on the ground (at more pay).

Another fast-growing area is in executive aircraft leasing. Rather than purchase an aircraft, many companies prefer to lease. The usual system is that the owner of the airplane agrees to provide and maintain the aircraft and also to provide the flight crews. The company that is going to use the aircraft agrees to use it for a certain minimum mileage each year at a basic rental fee. Additional mileage is paid for at a slightly lower rate. A pilot on constant call to a lease operator can expect to earn not less than $1,000 a month; those registered as standby pilots make less. An ATR-rated turbojet copilot on standby could expect to pick up $100 for a day's work. Without the ATR rating he'd pick up around $70. Many standby pilots have assignments with air taxi firms or act as relief pilots for corporations when the main crew is unavailable.

FIXED BASE OPERATORS

Other possibilities for would-be professional pilots include working for—or as—fixed base operators, as airplane salesmen, as instructors, or even as air taxi or third-level carrier pilots. Another area—somewhat demanding in motor skills—includes crop-dusting, sign-towing, and skywriting. And there's also test-piloting, although as far as civilian pilots are concerned, this is a rather limited field. Aviation journalism is another field, but it too is pretty much closed; apart from having fairly extensive experience and a good college background, it's important to have an attractive and understandable writing style. Or a good editor.

Fixed base operators (FBOs) may own an airport, may lease it from a city or county or even from another commercial firm, or may administer it. Some FBOs have dealerships in new aircraft, others trade in used airplanes, and some may do both. Several have their own flight-training schools, and with the advent of the Cessna, Beech, Grumman American, and Piper pilot-training programs—all worthy of your consideration—it seems likely that many more FBOs will be offering this service to the public. An FBO also frequently keeps a twin-engine aircraft for air taxi and charter work (for either passengers or freight), and some run aerial survey companies. A few specialize in modifying standard aircraft, either as suppliers or fitters of such modifications as the various STOL systems or in conversions, which improve the top speed performance of factory models.

130

Depending on the size of the operation, an FBO usually has vending machines for coffee and food—some run truly excellent small restaurants. He'll sell gas and oil, charts, and numerous accessories for aircraft. He may provide mechanical service or even facilities for the repair of radios and navigational equipment. By and large, most FBOs are very decent people who don't make as much money as they deserve for their efforts.

Becoming an FBO can be quite a risky venture if you don't know what you're doing. Airplane sales help, but with the big companies now demanding money down for new aircraft—in the good old days a sort of "sale or return" system operated—you will need rather more capital to get started. If you've never been in business before, it's a good idea to work for some other FBO and find out what it's all about from the ground floor up. Most FBOs are in it for their personal love of flying, and an enthusiast, even if not very bright, tends to meet their approval more easily than someone who views winged machinery as something attractive to look at on a balance sheet.

A good way to join an FBO is as a flying instructor. The pay won't be much, and the qualifications required may be high. Apart from having a reasonable idea of how to teach and the knowledge of the psychology involved in teaching people to fly, you'll also need a commercial certificate and a flight instructor rating. The better schools usually pay instructors a flat rate per month, depending on their experience. Others offer a base rate plus so much an hour for actual flight time. But a qualified aircraft mechanic would rate higher with a smaller FBO than a person who held just a flight instructor rating.

Another alternative is to instruct part-time, on weekends. Rates range from around $5 an hour upward, depending on the aircraft. There should be quite a rate rise here in the near future, for it is generally acknowledged that instructors are grossly underpaid. Skilled instructors, it is true, can make as much as $20,000 working with sophisticated machinery. Not too surprisingly, there aren't too many of these positions open.

Selling aircraft is a specialized business, and aircraft salesmen are rather more sophisticated than the average salesman. It is possibly more difficult than selling time on television, space in newspapers, or insurance. (It also demands nearly impeccable flying skills.) If you were buying an aircraft, you would by no means be indulging in an impulse purchase, even if you had just won the lottery. You've already weighed the pros and cons, know whether you can afford it, and have more or less made up your mind which

131

type of airplane is going to suit your budget and your intentions. A salesman is not going to be able to talk you into buying a newer or different model than the one you've already been considering. The best he can do is to appraise you of information you had not included in your calculations. The customer has a pretty good idea of the year, make, model, and (if it's a used airplane) the number of hours total and engine time he'd be willing to accept at a given price.

A person selling aircraft must be competent. In addition, he should have as good an idea of the competing model(s) as of those he is actually selling. In flight each maneuver from starting the engine to switching it off must be smooth and professional. There is no place here for fumbling for switches or for showoff flying. If the customer thinks the salesman/pilot is fudging the stall to give him a smooth ride and wants to take the controls—that's something else. But he shouldn't be left looking for his stomach after an especially vigorous power-on stall when he wasn't expecting it. If asked to demonstrate the airplane's capabilities, the salesman/pilot should explain slowly and carefully everything that is to be expected.

You yourself must also feel comfortable demonstrating stalls, spins (if approved), and various permitted aerobatics, such as lazy eights, chandelles, and so on. Even more, you must be able to demonstrate them in textbook style, but without embarrassing your possibly backhanded customer. According to those who sell airplanes for a living, the final decision to buy is made in flight. So you see, everything depends on how you fly that airplane.

Back on the ground a fair amount of time should be allowed for a discussion of price. It's usually possible to arrive at a price on which both can agree, though still too many salesmen work strictly

FIG. 5.8

Boeing-Boelkow BO-105C

from the bluebook and are little better in judging the quality of an aircraft than a car salesman the quality of a car. Besides, these days there's so much optional equipment involved in most aircraft, whether in radios or other equipment, that something can be worked out that suits both parties.

Typical salaries are $500 a month base, plus commission, which may range from 5 to 15 percent of the price of the airplane, or the base salary plus 10 percent of the gross profit. Sometimes no base salary is paid, and 25 percent of the net profit on new airplane sales is provided in its stead. Some manufacturers' representatives are paid a flat salary, say $12,000, and must meet a certain sales quota; earnings increase if the quota is exceeded.

One still uncommon area in aviation, sometimes associated with an FBO, is the flying club. Several pilots have found a home away from home in this field, either in setting up a flying club and then running it or in taking over floundering clubs, putting them into shape, and then working as club secretary. From the point of view of an FBO, setting up such a club can be quite a profitable under-taking as well as providing members with lower-cost flying. It is, however, important that the two operations should be separate, that is, that the club and the FBO should not share an identity. The club will want to own one airplane at least, which will require maintenance work. Several members may want to move up to own-ing their own aircraft, and provided the FBO has looked after them well as club members, it is not unlikely that they will purchase their next airplane from him. Finally, there's always the possibility of an arrangement with the club for the use of its machinery during the time when it is not in use, so that the FBO does not need to have too big an investment in rental aircraft of his own.

Crop-dusting, like most Ag piloting, is dangerous, not just be-cause there's a lot of low-level flying involved but because many of the chemicals used on crops these days are downright poisonous. The crop-duster can either be in business for himself, or he may work for a company that specializes in this field. Large farms usu-ally contract for the service on a long-term basis, and the security of income is thus usually quite good.

Apart from the chemicals used, the other drawback to crop-dusting is that it is seasonal. Usually there's a certain amount of traveling around to where the crops are coming in. The pilot can expect at most about six months' work a year. But at up to $25 an hour, quite a tidy sum can be put to one side during a six months' period. Then there's the question of equipment. The aircraft is magnificently bleak, with the appearance of being from some other

FIG. 5.9

The Rockwell Thrush Commander is the largest all-purpose agricultural aircraft in production today.

universe when compared with the sort of machinery found at major airports. Frequently it has an open cockpit, and with the numbers of telephone poles and powerlines strung across the country, life can be dangerous. Consider crop-dusting if your reflexes are faster than average, and have regular medical checkups.

Firefighting from the air is safer but not too different. Safety is relative, however, as firefighters may have to fly at night when such obstacles as telephone poles and unmarked high points may present themselves unseen.

Helicopter pilots are finding increasing demands for their services in agriculture and forestry, with new opportunities in such diverse fields as public health, traffic reporting, law enforcement, working with oil companies flying to offshore rigs, and pipeline laying and patrolling. Even the public utilities are using helicopters for maintenance. Still, many of these posts are being filled by ex-military helicopter pilots.

Finally, there's sports flying. Air shows frequently provide cash for race winners as well as trophies, although usually the amount one wins in a year will not cover his expenses. Aerobatic pilots—provided they are fine performers—can make a living, but the work is arduous. And so are the logistics—aerobatic pilots must crisscross the country, which usually means long hours of flight between demonstrations (the types of airplanes they use are neither very fast nor very long-range). However, top pilots can make money—$100 a minute is not exceptional for the very best. Their aircraft must be in top condition at all times, and it really isn't much fun living out of a small suitcase for days at a time during the summer months.

FAA LIST OF CERTIFICATES, RATINGS, AND TYPE RATINGS

STUDENT

PRIVATE
COMMERCIAL
A.T.R.

Airplane—Single-engine land
Multi-engine land
Single-engine sea
Multi-engine sea
Rotorcraft—Helicopter
Gyroplane
Glider
Lighter-than-air—Airship
Free balloon
Large aircraft—type rating
Turbo-jet powered aircraft—type rating

Instrument
Instructor
Instructor-
Instrument

Non-pilot ratings

Ground instructor
Flight engineer
Flight navigator
Air-traffic-control tower operator
Aircraft dispatcher
Mechanic—Airframe
Powerplant
Airframe & powerplant (A&P)
Repairman
Parachute rigger

PART IV

6

A Look at Some Popular Two-Seat Trainers

Specialized trainer aircraft really came into their own during World War II. They were designed to have flight characteristics similar to the war machines that the neophytes would soon fly but without the speed and other critical characteristics of the genuine article. Today's trainers are built primarily for training, although various options are available that make them acceptable as cross-country machines in miniature. The design parameters of most successful trainers include a rugged ability to withstand less than expert handling, ease of maintenance, and reasonable value in terms of purchase price and running costs.

CESSNA 150

The Cessna 150 is a two-seat, high-wing monoplane powered by a 100-hp Continental engine. Originally produced in 1958, it marked the reentry of the Cessna company into the two-seat aircraft market, following the demise of their 120/140 series some years earlier. As a trainer it has provided well over half the present generation of today's pilots with their initial flight training, along with much fervent argument on the relative merits of the high-wing versus low-wing aircraft.

With the 150 and its cousin the 172, Cessna has established the longest and best-selling line of aircraft ever produced, with more than 20,000 sold. You'll find 150s all over Europe, in tweedy green fields in England, at exclusive resorts on the Mediterranean, in South America, and in Australia. It has a remarkable range when

FIG. 6.1

Cessna Model 150 in aerobatic disguise

the (optional) long-range tanks are fitted and will travel nearly 900 miles, albeit rather slowly. It is the staple of more than half the training establishments around the world.

By 1964 the earlier models had given way to the Model D, featuring the then new wraparound rear windshield and an additional 100 pounds in gross weight. Its successor, Model E, had a redesigned instrument panel, and the Model F introduced a swept tail. Cessna offers four models based on the 150 in this country, including an aerobatic version. (A fifth version, produced in France by Reims Aviation S.A., is identical, except that it is powered by a O-200-A engine manufactured by Rolls-Royce under license.) All models have a gross weight of 1,600 pounds, and their useful load varies from 535 to 620 pounds. The Commuter, intended for basic transport, is equipped with numerous items (including radio) that are only optional in other versions.

Flight characteristics are just a bit dull. The control inputs are carefully slow and balanced and for learners absolutely safe— which is probably why flying schools buy the model year after year. On takeoff you raise the nosewheel at an indicated 55 mph and climb out between 70 and 80 mph. Let the nose down a fraction and you can cruise climb at 85 to 90 mph and be up and on your way quite briskly. Cruise speed is around 115-mph, not exactly the greyhound of the skies, but a good deal faster than traveling by car.

At low speeds you have plenty of aileron control right into the stall. The 150 can be made to spin if you want, and correction is easy. Just apply full opposite rudder to the direction of rotation, then ease the yoke forward to about neutral. As rotation stops, neutralize rudder and make a smooth recovery from the dive. It's as easy as ABC, another reason why flight instructors like it so much.

For landings, power-on or power-off approaches are usually made at 70 to 80 mph with the flaps up or at 60 to 70 mph with the flaps down. For really short-field performance, use full flap deployment (40 degrees of flap) and bring the aircraft in with plenty of power at 60 mph. The actual touchdown speed can be very low—less than the 50 mph indicated airspeed if you do it properly.

Stalls are very gentle and, unless provoked, register as nothing but a gentle mush with a high rate of sink. In the stall the rudder is adequate to hold up a wing that wants to drop. Stalls in level flight occur at 55 mph corrected air speed (the indicator will read just over 50 mph), at 49 mph with twenty degrees of flap, and at 48 mph with forty degrees of flap.

The aerobatic version is not intended for competition but for training. It is nicely balanced and will loop and roll and let you enjoy putting the green where the blue is normally found. You won't learn much about inverted flight, as the fuel system is limited to about ten or fifteen seconds of fuel flow while inverted. But it is a fine machine to begin on.

As transport it's a bit slow compared to other airplanes, and it has only two seats. On the other hand, it will carry you across country a good deal quicker (and safer) than a car on the highways, possibly a bit cheaper too. As a first airplane to own there is much to commend the 150. Older models are available for less than $3,000; newer versions cost a bit more. A four- or five-year-old airplane, depending on condition and equipment, will begin at $3,500 or slightly more. The fuel used is 80-octane, cheaper than 100/130 octane; you'll use 5½ to 9 gallons per hour, depending on your speed.

If you are buying a Cessna 150 that has been used as a trainer, get a friendly A&P (airframe and power plant) mechanic to check it thoroughly, especially the landing gear. It's all mendable, but some caution in advance will ensure that you're not the one to foot the bill.

Specifications

Gross weight	1,600 lbs.
Empty weight	1,000 lbs.
Useful load	600 lbs.
Wingspan	33 ft. 4 in.
Wing loading	10.2 lbs./sq. ft.
Length	23 ft. 9 in.
Height	8 ft.
Fuel capacity	26 gals. (80/87 octane)
Range (cruising at 75 percent power)	475 miles (standard tanks)
(long range, optimum)	565 miles (standard tanks)

Powerplant

Engine	Continental 0-200-A rated 100 hp at 2,750 rpm
Propeller	Fixed pitch 69 in. diameter

Performance

Takeoff run	735 ft.
Takeoff over 50-ft. obstacle	1,385 ft.
Rate of climb	670 fpm
Service ceiling	12,650 ft.
Maximum speed	122 mph
Cruise speed (75 percent power at 7,000 ft.)	117 mph
Stall speed, flaps up, power off	55 mph
Stall speed, flaps down, power off	48 mph
Landing roll	445 ft.
Landing over 50-ft. obstacle	1,075 ft.

PIPER 140

Long before the Cessna 150 was even so much as a twinkle in the Cessna designers' eyes, a man named Taylor designed a two-seater

FIG. 6.2

Piper Cherokee 140

aircraft destined to become a household word. Mr. Taylor's airplane was, of course, the Piper Cub; to this day it remains in production as the Super Cub for those who prefer their airplanes built in the good old-fashioned way.

The Piper Cub was for aviation what Ford's Model A was for car owners. It was the prime means of training pilots during the late 1930s until well into the war years. Military (training) versions were made of it; artillery spotting airplanes were another version. Finally, it gave birth to the Pacer, which was developed into the Tri-Pacer by enlarging the tail wheel and moving it up to the front. Then came the Cessna 150. Sales at the Piper factory sagged.

Originally introduced in 1960, the Cherokee family, of which Piper's 140 is the junior member, is now quite large, ranging from the original 150-hp version to the Big Six (capable of lifting its own weight in useful load and consequently fashionable in the backcountry as a mini-freighter). There are retractable landing gear versions, too, sleek with singular performance.

But the 140—and the Flite Liner version, made especially for flying schools—is the mainstay of the line and is the principal alternative as a trainer to the Cessna 150. The basic machine is a two-seater, with an option to add two more seats at the rear. It has a low wing, is faster, and has a greater useful load than the 150.

As you walk around the aircraft, it seems to be very low to the ground. The main wheels are set widely apart—10 feet apart, some thirty inches wider than the 150, and one-third of the wingspan of the 140s. This spells good crosswind landing characteristics. The low wing enhances the compression of air (called *ground effect*) when an airplane is close to the ground, making it easier to touch down gently almost every time you land. In fairness to the 150, it should also be added that some students find it very difficult to make the best use of this and take longer to avoid ballooning when they land. (See Figure 3.4c.)

There's only one door, which leads into a quite handsome cockpit. New 140s have air conditioning as an option. The instrument panel is well laid out, with everything falling into place in the standard T-pattern. In the Flite Liner version you'll discover that there are no toe brakes but only a center-mounted "Johnson bar," which goes back to the days of the Tri-Pacer. And for student training, you fill only to the eighteen-gallon standpipe marker, although for cross-country travel you can top each tank with twenty-five gallons of 80/87 octane.

The control pressures of the 140 are heavier than the Cessna 150 and require more use of trim. But the aircraft is highly stable and a pleasure to fly. It is sensitive in pitch, as you'll discover when you use the flaps: There's a noticeable change in attitude, but once you trim out the control pressures it's really quite easy. The 140 is probably a little more difficult to land well than the 150, but it seems faster in all modes of flight. For takeoff you also must work a little bit harder to ease it off the ground.

The original 140 became a trainer in 1964 when it was powered by a 140-hp Lycoming. That model was known as the PA-28-140, but today's version uses Lycoming's 0-320-E2A engine of 150 hp. Stalls are gentle and occur at about 55 mph; normal cruise is about 132 mph. Maximum speed is just short of 140 mph, and it's placarded against spins. If you manage one inadvertently you'll discover that after a turn and a half you're in a spiral dive and approaching the redline speed too rapidly for comfort. As a cross-country airplane, the 140 is very comfortable indeed.

The newer Flite Liner is the first light aircraft to have a progressive maintenance scheme. Progressive Maintenance, which was invented by the military and is used by the airlines, means that every few hours a number of checks are performed on an airplane, so that—in the case of an airline for example—the airplane is only out of service for two hours in every twenty-four. Obviously, this is a lot more economical for an airline than having to take an airplane out for a whole day once every six days, as scheduling can ensure that an airplane is being maintained out of peak hours.

For the Flite Liner it works like this: Every fifty hours the aircraft is taken out of service for a series of brief checks and for lubrication. Four of these fifty-hour checks are the equivalent to the FAA's 100 hour inspection. In effect, this means that there are 200 hours between "100-hour inspections," although the airplane is inspected twice as often as it would be under the older schedule. This system saves money on maintenance and keeps the airplane flying when otherwise it might be taken out of normal service.

Older 140s represent good value in the secondhand market, at around $3,250, with prices ranging upward depending on condition and radios. A problem to look for, especially if the aircraft hasn't been well cared for, is leakage from the fuel tanks. New 140s run from just over $11,500 basic to $14,060 for the Flite Liner—some $2,000 more than an equivalent Cessna 150.

Specifications

Gross weight	2,150 lbs.*
Empty weight	1,237 lbs.
Useful load	913 lbs.
	(560 lbs. for the Flite Liner as a trainer)
Wingspan	30 ft.
Length	23 ft. 3 in.
Height	7 ft. 3 in.
Fuel capacity	50 gals. (80/87 octane)
Range (cruising at 75 percent power)	634 miles
(long range, optimum)	790 miles

Powerplant

Engine	Lycoming 0-320-E2A rated at 150 hp
Propeller	Fixed pitch

Performance

Takeoff (over 50 ft.) obstacle	1,700 ft.
Rate of climb	660 fpm
Service ceiling	14,300 ft.
Maximum speed	139 mph
Cruise speed (75 percent at 6,500 ft.)	132 mph
Stall speed, flaps up	61 mph
Stall speed, flaps down	55 mph
Landing (over 50 ft.)	890 ft.

* Although 2,150 lbs. is also the gross weight of the Flite Liner, its normal operating weight is 1,800 lbs.

145

FIG. 6.3

Grumman American Trainer

GRUMMAN AMERICAN TRAINER

No matter what the Grumman-American Aviation salespeople say, the American Trainer really ought to be called the Yankee Trainer, for despite the occasional disparaging connotation of the term "Yankee," the airplane is, first, the product of a northern state. Second, it has been daringly aimed at a market that has been successfully serviced for many years by two of the largest manufacturers in the business. And third, it has a radically different philosophy of what a training airplane is all about.

In distinct contrast to the two preceding aircraft, the American Trainer is very much a pilot's airplane. To be sure, it is admirably docile if you fly it right, but it does not sit back and turn the other cheek if you goof up. It is modern in design, though a bit chunky in appearance, having been created on the drawing board of designer extraordinary Jim Bede. (Bede, you may remember, broke the distance record for light aircraft in his U2-type creation and is now putting on the market one of the fastest and most revolutionary single-seaters for do-it-yourselfers, the BD-5.)

It offers the new pilot considerable economy of operation and, at $9,000, relatively low cost for a new airplane. In achieving these

146

objectives, use has been made of such space-age technology as metal-to-metal bonding. Aluminum honeycomb provides strength and rigidity at low weight. And there are no fuel tanks as such. The wing spar, a hollow tube around which the wings are built, doubles as the fuel tank.

The cockpit is a bit small for a tall pilot, especially after a long cross-country flight, but for anyone less than six feet tall, it is very comfortable, with a seat that adjusts upward when you bring it forward. The wingspan is only 24½ feet; the fuselage measures 19 feet 3 inches from stem to stern. But it looks tough—and it is. The landing gear has to be one of the most bump-absorbent on the market. It's made of eighty-five strips of fiberglass that have been laminated together, resulting in considerable strength and lightness. The track is over 8 feet wide, which makes for good ground stability.

Until the American Trainer made its appearance on the market, training aircraft had been designed to provide slow but positive responses to control inputs and inherently stable characteristics with a relatively slow rate of roll. Both the Cessna 150 and the Piper 140 are devoid of tricks and vices and are very easy for the student to pilot.

Then along came the American Yankee (from which the Trainer developed). It was advertised as both a training machine and a transport. It was modern and very fast and demanded an extra measure of alertness from its pilot. The Yankee's biggest problem was the relatively high stalling speed—65 mph. Although it was immensely controllable, some instructors felt it possessed a liveliness undesirable for beginners.

The engineers went back to their slide rules and came up with a new wing, changing the laminar flow wing of the Yankee by drooping the leading edge. This lowered the stall speed by 6 mph, and suddenly the new Trainer was in business. They sold 150 of them in the first five months of production and many, many more since. It seems that those instructors who felt the Yankee was too hot had second thoughts. Some apparently felt that maybe flying shouldn't be too easy, and this type of aircraft might help weed out those who should never have attempted to try to fly in the first place. It is truly a fun airplane. It has excellent control response and feels like a thoroughbred. Unfortunately, it is not presently rated for aerobatics.

The cockpit has a large plexiglass dome (tinted at the top), which slides back to let you in. You can fly with it cracked open at speeds below 120 mph. You climb up onto the wing, then dex-

147

terously, with your toe, ease up the seat cushion and clamber in. As you settle in your seat you notice the panel is well laid out, with everything immediately and logically at hand. What is best, though, is the visibility, for although you cannot see directly behind you, you have an excellent view in almost every other direction.

For taxiing, there's the castoring nosewheel. Once you get the hang of it (you steer with dabs of brake and rudder), you'll find you can spin the airplane around almost within its own length. You have the benefits of tricycle gear with the maneuverability of a tail-dragger. And you don't need terrific technique to handle it like a pro.

Normal takeoff is at the same speeds as other aircraft, though climbout is a bit faster. Rotate the nosewheel at about 65 mph and climb out at about 90 to 95 mph. Short-field performance is not outstanding at gross, although if there are no obstacles ahead you can lift the airplane off and fly it in ground effect air cushion till the speed builds up. Best angle of climb is 75 mph at sea level; best rate of climb is 90 mph.

An auxiliary fuel pump is switched on for takeoffs and landings and provides a backup to the engine-driven pump. The fuel system is one of the simplest, with sight tubes inside the cockpit (as on the French Rallye series of aircraft and the Piper Super Cub) showing exactly how much fuel remains.

Cross-country cruise speed, with the climb propeller, is 125 mph. If you fit the cruise prop, you'll add 10 mph to your speed and about 100 feet more to your takeoff roll. You'll also lose out on your rate of climb, which will drop from around 765 fpm to 720 fpm.

If you learn to fly in an American Trainer, you may take a bit longer to get there, but you'll learn good flying habits right from the start. And from an instructor's point of view, it certainly seems his needs have been carefully considered. A pilot with few hours of flight time who starts out in the Trainer should have little or no difficulties making the transition to larger aircraft. And a single-engine pilot wanting to go to multi-engines would find that transition easier, because of the American Trainer's slightly nose-down attitude in level flight. You need light hands to fly it well, but it makes you want to succeed.

Normal range is just over 400 miles at 75 per cent power; optimum range with no reserve at 10,000 feet altitude is 500 miles. Fuel consumption works out at about 20 mpg at a cruise speed of 125 mph, which is better than your favorite station wagon—which is why Big Oil doesn't care for them.

Specifications

Gross weight	1,500 lbs.
Empty weight	1,007 lbs.
Useful load	493 lbs.
Wingspan	24.46 ft.
Wing loading	14.9 lbs./sq. ft.
Length	19.24 ft.
Height	7.6 ft.
Fuel capacity	24 gals. (22 usable)
Range (cruising at 75 percent)	440 miles
(long range, optimum)	500 miles

Powerplant

Engine	Lycoming 0-235-C2C rated 108 hp at 2,600 rpm
Propeller	McCauley fixed pitch 71 in. diameter

Performance

Takeoff run	725 ft.
Takeoff over 50-ft. obstacle	1,400 ft.
Rate of climb	765 fpm (climb prop)
Service ceiling	13,750 ft.
Maximum speed	138 mph (climb prop)
Cruise speed (75 percent power)	125 mph (climb prop)
Stall speed, flaps up	61 mph
Stall speed, flaps down	59 mph
Landing roll	395 ft.
Landing over 50-ft. obstacle	1,065 ft.

BELLANCA CHAMPION CITABRIA

Citabria pilots tend to refer to conventional airplanes as "milk stools," think of other pilots as "airplane drivers," and wonder when they will ever learn about the real joys of flight. Occasionally, when one of these "airplane drivers" visits the bush country, he is invited to partake in a traditional ceremony. Nurtured during his formative years in tricycle-geared machinery, he finds himself in the cockpit of this seemingly uncomfortable and venerable airplane. The walls

149

FIG. 6.4

The Decathlon is the super aerobat of the Citabria family.

are of fabric, and the whole thing looks as if it is put together with string. This rite of manhood is compounded by the fact that there's no control yoke (only a simple stick), you can't see much forward when you're on the ground in the front seat (you can see nothing if you are the instructor in the back), the heel brakes are tiny buttons on the floor, and the wind howls in around the door. The problem is made even worse by the fact that the seat is not adjustable and you must sit on a parachute.

Surprisingly, once a pilot gets used to the machine he finds it a joy. It's approved for loops, rolls, tailspins, and the like. Although noisy, it's an excellent airplane for the student, as he must learn how to coordinate the stick with the throttle and rudder to make maneuvers properly. Finally, with those heel brakes you learn a dexterity of foot in keeping the machine from veering off into the greenery at the edge of the runway every time the wind is slightly askew. You have to land it right—and some instructors feel that that's a good thing too.

The Citabria's name, as most people know by now, is "airbatic" spelled backward. It is, in fact, a beefed-up version of the old Aeronca 7AC two-seater, which was bought by Champion. A modern, nonaerobatic version of the aircraft is marketed at less than $8,000 (new) under the name Bellanca 7AC Champ. Its virtually the same airplane but with a different engine and of course without the licensing to do aerobatics.

The Citabria comes in a variety of versions, all fabric covered. The originals came with a 65-hp powerplant, but Citabrias are

150

found with 100-, 115-, or 150-hp engines. At cruise speed there's a difference of some 25 mph, but the bigger-engine birds show their stuff on takeoff with STOL performances. Other variations include fuel injection on the 150-hp model, a fuel header tank for inverted flight, and an inverted flight oil system. The heel brakes are perhaps the worst feature, as the ones at the front seem to have been carefully placed so that when you have full rudder and need to brake, as you might in a crosswind, you simply cannot reach them in normal-size shoes without letting up on the rudder. The instructor's brakes at the back, needless to say, work fine.

As an aerobatic performer, the Citabria will do most maneuvers well enough, although you will build up an enviable set of muscles in the process. As an added safety factor, it is virtually impossible to reach the redline speed if you're the right way up. According to the pundits of the backwoods airports, the only way to reach redline is inverted in a sixty-degree dive.

Consider learning in a Citabria. You must be smart to fly one. You'll have to learn good flying technique from the start. A number of flying schools keep a Citabria or two to use in courses for advanced students, but many instructors still swear by them for beginners.

Prices range from around $3,000 to as much as $8,000 for fuel-injected models in the secondhand market. It's not the fastest machine for the money, but it will take care of you. If you want to be more than just an airplane driver, you could do worse than the Citabria.

Specifications

Gross weight	1,650 lbs.
Empty weight	1,107 lbs.
Useful load	543 lbs.
Wingspan	33 ft. 5 in.
Length	22 ft. 7 in.
Height	7 ft.
Fuel capacity	39.5 gals.
Range	520 miles

Powerplant

Engine	Lycoming 0-320-E2A rated at 150 hp
Propeller	Fixed pitch

Performance

Takeoff over 50-ft. obstacle	630 ft.
Rate of climb	1,120 fpm
Service ceiling	17,000 ft.
Maximum speed	130 mph
Cruise speed (75 percent power)	125 mph
Stall speed*	51 mph
Landing over 50-ft. obstacle	755 ft.

MR. PIPER'S CUB

Depending on your point of view, the J-3 Cub was wonderful or terrible. (Actually, it never really was Piper's Cub at all, since—as mentioned earlier—it was designed by C. G. Taylor, who was formerly president and chief engineer of the Taylor Aircraft Company, better known today as the Piper Aircraft Corporation.)

The Piper Cub is a high-wing monoplane with a 35-foot wingspan and a length of 22½ feet. It was a linen-covered, steel-tube–framed product before it was discontinued in 1947. It was reborn a Super Cub with today's 150 hp and a fancy price tag of more than $13,000. The J-3 Cub was powered by three different types of engine—Continental, Franklin, and Lycoming—all of which were rated at 65 hp. The Continental version was by far the most common.

Most J-3 Cubs flying today have a fuselage of welded steel tubing covered with fabric. The wings, also fabric covered, have metal or wooden spars. The gas tank holds twelve gallons, and the range is 200 miles, depending on which way the wind is blowing. Cruise speed is 85 mph; the aircraft will stall at speeds in the low 30s. It has a semi-steerable tailwheel, and the shock absorption of the landing gear comes from rubber bungee cords and low-pressure tires.

The modern variant, the Super Cub, will almost lift its own empty weight in useful load, cruises at 115 mph, and gives STOL takeoffs and landings time after time.

Like the Citabria, the Cub is a tandem two-seater. To an owner, a Cub is not merely comfortable but it is a thoroughly dependable machine. If you think you might like to try this type of flying, it's been said that if you can fly a Cub you can transfer to almost any other airplane with ease.

The Cub's appearance tells its story, although some owners clip

* The Citabria is not fitted with flaps.

FIG. 6.5

Super Cub

the wings to improve the roll rate. It isn't really meant for long-distance travel. Its purpose is levitation, exploring the inner reaches of the sky, discovering what flying is all about. It won't whisk you to Palm Beach or Acapulco, and if that's what you want it for, you really should try something else.

The Cub is not an expensive airplane to maintain or to run. Even the Super Cub is inexpensive, costing around $9.00 an hour for 200 hours' use per year. And today you can recover Cubs in permanent fiberglass fabric, which gives them a skin life comparable to an all-metal aircraft.

Depending on the aircraft's condition, expect to pay $1,500 to $5,000 and to spend up to a year looking for one. Good ones are scarce, and your best bet may be to find one that's being sold cheap and then rebuild it yourself, which is one of the best ways to learn all about flight.

Specifications*

Gross weight	1,220 lbs.
Empty weight	680 lbs.
Useful load	540 lbs.
Wingspan	35 ft.
Length	22 ft. 6 in.
Height	6 ft. 7 in.
Fuel capacity	12 gals.
Range	200 miles (approximate)

* Figures quoted are approximate, since J-3 Cub performances tend to be highly idiosyncratic.

Powerplant

Engine	Varies—see text; rated at 65 hp
Propeller	Laminated wood, fixed pitch

Performance

Takeoff over 50-ft. obstacle	600 ft. (approximate)
Rate of climb	500 fpm
Service ceiling	You'll be out of gas before you find it
Maximum speed	85 mph
Cruise speed (75 percent power)	75 mph
Stall speed	35 mph
Landing over 50-ft. obstacle	500 ft. (approximate)

BEECHCRAFT SPORT

In much the same way that there are automobiles and Rolls-Royces (or Mercedes-Benz, depending on your prejudice), so, it is said, there are airplanes and Beechcrafts. Instructors and students both enjoy this Rolls-Royce of training airplanes, as one of its best features is that at cruise speed you don't have to shout. It is also aerobatic, being approved for rolls, Immelmann turns, loops, spins, chandelles, and limited inverted flight.

This surprises a number of people, whose view of the Beechcraft line is that it caters foremost to the carriage trade. Like Rolls-Royce, the Beechcraft line is not exactly famous for its innovative engineering, relying on the perfection of existing practice for its appeal in the marketplace. And it has worked most successfully. Beechcraft owners slip to the top of the line when it comes to refueling, and their pilots are accustomed to having the best of service wherever they travel. Indeed, it is said of the organization of Flying Physicians that there are only two kinds of members: those who own Beech Bonanzas and those who wish they could. Still, it has to be paid for.

It is not surprising, therefore, to find that when Beechcraft decided to market a training vehicle, they put their expertise into marketing the best that money could buy. Consequently, the Beechcraft is more expensive than other trainers, but needless to say it offers proportionately more.

A relative newcomer to the field of training ships, the company expects to tap a new source of Beechcraft owners from the training field. Both Cessna and Piper have long known of the product loy-

FIG. 6.6

Beechcraft Sport

alty engendered by kindly training ships, and the more upwardly mobile pilots tend, by and large, to stay with the company in whose product they began to fly. Consequently, the Sport is the mainstay of Beechcraft's own integrated flight-training system. In addition to its excellent forward visibility, the Sport has four seats in standard configuration, cruises 130 mph plus, for a range of nearly 800 miles.

Although not as vigorous an aerobat as its larger-engine brother, the Sundowner, the Sport rolls to the manner born. Its loops are comfortable 2 to 2½ times gravity, and it snap rolls with vigor. Spins can be tricky if you don't hold the control column fully back. Perhaps the Sport's only drawback is the way it picks up speed.

The instrument panel is exceptionally well designed—clear and easily understood. The trim wheel, as on the Grumman American Trainer, is positioned between the seats. Driving the prop is that favorite Lycoming 150-hp 0-320 series engine.

The Beechcraft Sport lopes off the ground at around 60 to 65 mph, and you don't need to muscle it off the deck. It wants to fly, so ease back a tad to lighten the weight on the nosewheel, and the Sport will fly off on its own.

Stalls are quite normal and gentle, although there does not seem to be quite as much prestall buffeting as in other trainers. Landings are simple, especially if you don't mind making the effort to fly by the numbers. And that, many instructors maintain, is excellent practice for those who intend to move up to retractable and twin-engine aircraft, such as the Beechcraft Sierra (a retractable version of the Sport) or the twin-engine Bonanza. The technique seems to be to hold 80 mph with one notch of flap and about 1,700 rpm from the last half of your downwind, slowing to 70 mph with full flap for final. Featherlight touchdowns time after time are almost guaranteed.

List price of the Beech Sport is $18,750 without radios or gyro instruments. For the money you get an aircraft with a direct hourly operating cost of $6.61, according to Beechcraft, based on using 7.8 gallons per hour of 80-octane gas at 45 cents a gallon. Several options are available. The aerobatic, which has a G-meter, inertia reel shoulder harnesses, and quick-release doors, costs $2,000 more. Avionics (radios) go from a basic nav-com system at $1,725 to a full package including autopilot at more than $8,000. To go with that there's the professional option, which gives you all the instrumentation you'll ever need, at around $2,500.

The only problem with the Sport is that if you do fit it out completely, you'll have to forget about those two in the back, in which case you'd be rather better off with its bigger-engine brother for just $3,000 more. You get 200 pounds more of useful load.

Specifications

Gross weight	2,250 lbs.
Empty weight	1,374 lbs.
Useful load	876 lbs.
Wingspan	32 ft. 9 in.
Wing loading	15.14 lbs./sq. ft.
Length	25 ft. 8½ in.
Height	8 ft. 3 in.
Fuel capacity	60 gals.
Range (cruising at 75 percent power)	767 miles
(economy, 55 percent at 10,000 ft.)	883 miles

Powerplant

Engine	Lycoming 0-320-E2C 150 hp at 2,700 rpm
Propeller	Fixed pitch

Performance

Takeoff run	885 ft.
Takeoff over 50-ft. obstacle	1,320 ft.
Rate of climb	700 fpm
Service ceiling	11,100 ft.
Maximum speed	140 mph
Cruise speed (75 percent power at 7,000 ft.)	131 mph
Stall speed (landing configuration)	56 mph
Landing roll	590 ft.
Landing over 50-ft. obstacle	1,220 ft.

7

Airplanes from Kits

The celebrated author of *Jonathan Livingston Seagull*, Richard Bach, himself a noted pilot, once suggested that all who wish to take to the skies should first serve an apprenticeship that would include building the airplane they will ultimately fly. The idea has merit. Certain mariners would not undertake a long voyage without the surety of their own handiwork. And today there is a burgeoning number of pilots who build their own airplanes. At last count more than 5,000 "homebuilts," as they're called, had taken to the air. It would not be an exaggeration to say that some of these represent the art of aircraft building at its summit.

Some people build their own aircraft because they want to race. Some claim that it is the least expensive way to own an airplane— and it can be, if you have some basic knowledge before you start. Others will tell you that it is the only way to enjoy a truly customized vehicle—and they have a point, too. Still others prefer to build rather than to buy because they have intentions that cannot be met with conventional aircraft. There are very nearly as many reasons for homebuilts as there are varieties of airplanes. But before you think seriously of building your own airplane, you ought to join the Experimental Aircraft Association (EAA), founded in 1953 almost by accident.

After World II there was much optimistic talk about every family having an airplane or helicopter in the garage. There were also many veteran pilots around. But the economic truths were quickly

158

perceived: buying an airplane was an expensive proposition. At that time there was much less surplus cash available than today, and while the prices seem low by comparison with those now current, our nostalgia is misleading.

The result was that many individuals began to build their own aircraft. It was almost an underground movement, and it wasn't until 1953 that some twenty of these builders around the country got together to form the EAA. The concept was that they could swap ideas, let others know where inexpensive parts could be obtained, and exchange thoughts and opinions on the best ways of designing and building airplanes.

Today the EAA has more than 50,000 members around the world. It publishes a number of books of great value to homebuilders, and its members have put more than 4,000 privately built aircraft into the air. As of this writing, there are more than 7,000 airplanes under construction by EAA members.

You can write for details about the EAA and its activities to Box 229, Hales Corners, Wisconsin 53130. Annual membership is $15.00, which includes a subscription to *Sport Flying* magazine, a publication of the association. If you're tempted to build, you can have access to a welter of information from other members, many of whom had chosen this route to aircraft ownership.

As a rule, approximately 3,000 hours and $3,000 are needed for the construction of a homebuilt. Many knowledgeable homebuilders feel that a beginner in the field will be better able to cope with the problems of building an airplane once he's had some experience with a job that can be concluded more quickly. Repairing or overhauling an existing airplane is usually easier and takes less time, especially if you've never done it before.

Despite the fact that flying is the safest form of locomotion, according to National Safety Board statistics, government regulation of flying and of pilots and aircraft has increased to the point where such regulation employs more than one person for every aircraft in the nation.

Airplanes produced by regular manufacturers have to go through extensive testing before being granted certification. Once granted, it is extremely difficult to make major changes in the design to improve it, which requires recertification. For the homebuilder, however, there is a special category known as "Experimental Aircraft" that allows the private constructor to acquire a Special Airworthiness Certificate. In order to take advantage of this procedure (details of which are covered in FAR Pt. 21.191 and 21.193), the would-be homebuilder should first decide on which design he wants and then check with the supplier of the kit to see that the kit does satisfy the regulations. The key requirement is that 51 percent of the kit must be built by the purchaser. A telephone call to the local Engineering Manufacturing District Office (EMDO)—usually found at a General Aviation District Office (GADO)—at the FAA will confirm whether the kit is acceptable, and once the kit has arrived the FAA will usually send an inspector over to check the kit and to give some useful hints on construction. If you're not very good he may get in your way.

Depending on work load at the local EMDO offices, at least two and usually three inspections are made during the period of construction. For example, the inspector may want to check your control mechanisms before you close off the area, and there will be a final inspection before the airplane is permitted to fly. Depending on whether your engine and propeller are certificated or not, you will normally be given a period of fifty hours in which the airplane is given a shakedown; during this time poor design, faulty construction, and poor flying habits are expected to show up and be corrected. The FAA is interested in stable flying characteristics, and through a series of weight and balance tests, checks the center of gravity.

During the fifty-hour period you will be limited to a flight-test area, which normally is not more than twenty-five miles in radius from your home base. Upon satisfactory completion of the shakedown, a new certificate is issued that is valid for one year. Once granted the certificate, you must not make changes in the design; if you do, you'll be put back into the earlier classification, which places several restrictions on you. If you use an uncertificated engine, your shakedown period might be extended to seventy-five hours; it is entirely at the discretion of the FAA. But once they're satisfied that you've built a decent airplane, they relax the restrictions and you can then fly where you will. You will never be able to fly passengers for hire in your airplane, but you can fly them for fun. The homebuilder-pilot must hold at least a student-pilot's license and a third-class medical.

If this sounds like a lot of official interference, don't be put off. The people at the FAA who supervise this program are mostly pilots themselves, and many have had personal experience building aircraft. You may find that the FAA official assigned to your building project takes almost as much interest in your craft as you do.

Anyone reasonably skilled will have no difficulty in fabricating an aircraft from plans. Most designers these days provide comprehensive Bills of Materials, with helpful recommendations about where parts may be obtained. Designer Ladislao Pazmany, whose PL-2 is the only amateur-built airplane currently manufactured as a military trainer in four countries, is not merely an excellent designer and draughtsman but also a first-rate engineer and writer. His book, *Light Airplane Construction*, is something of a classic among the homebuilding fraternity and is well worth the $7.00 Pazmany charges. The PL-2, a smart side-by-side two-seater, takes about 3,500 hours to complete from scratch. However if you make use of prefabricated parts, Pazmany claims you can probably save as much as 1,000 hours in construction.

A popular source of information on where parts may be found is *Trade-a-Plane*, a newspaper published three times a month in Crossville, Tennessee. Although it primarily lists aircraft of all kinds, shapes, and sizes for sale, it carries extensive advertising of parts. If you need a crown-wheel for your widget gunion, you may not find it listed under widget gunions but you'll find it somewhere in *Trade-a-Plane*.

The homebuilder who wants to get flying with the least hassle of all has another method—building from a kit. As stated before, the only stipulation is that 51 percent of the work must be completed by the builder. In all the better kits available on the market today, all the most difficult work—fine machining, for example—has been done. Although there may still be some cutting and fitting in the assembly, most can be completed by the average handyman.

Another advantage in building from a kit is that you can make substantial savings in time. Some aircraft kits take only 600 hours. But experienced builders warn that it is not the airframe that takes time but the fitting of incidental items. Instruments and electrical wiring can be headaches, as can control-system hookup and adjustment, cabin heating and cooling, and even the Pitot system. Still, these are always problems. The secret is to set aside a couple of hours each evening, plus five or six hours each weekend, and to stay with this schedule come hell or high water. That way you'll be amazed at how quickly your airplane is put together. And if you plan carefully before you start each section, you'll save additional time.

Some kit manufacturers will send your aircraft section by section, so that you do not need to buy the whole thing all at once. Jim Bede, one of the doyen of homebuilt aircraft, supplies parts for his BD-4 (the most popular American homebuilt until his BD-5 came on the scene) in no less than seven materials packages, none of which costs more than $750.00. Of course, if you buy the lot at one time, there is a saving.

161

For your first venture in building it is advisable to follow the manufacturer's recommendations pretty closely. Modifications to customize your airplane can sometimes be tricky, and unless you have a friend who is an expert in the field, it is usually not worth the bother. You can save more money knowing where certain parts may be available cheaply. This is where the combined experience of members of the EAA is so helpful—the grapevine works well. But the better kits on the market are fairly complete, and you have the additional satisfaction of knowing that all the parts are of aircraft quality and that thorough checks have been made to ensure their reliability.

If you're going to build from the kit, you probably have all the tools you'll need in your workshop. Normal household hand tools plus an electric drill and a variable-speed saber saw are usually all that are required. You can make life much easier for yourself if you construct a large, perfectly flat worktable on which to assemble some of the parts.

Finally, before you decide to go ahead, consider what you'll be doing with your aircraft when it's completed. Do you want something to fly around in on weekends, perhaps to hop over to the next town? Or are you intending to pack your wife and children in with you and set out to explore the wide open spaces? Possibly you may want to fly aerobatics. Whatever your needs, think carefully about what you will want the airplane to do before you commit yourself to the project. There is a good selection on the market, and although it can be expensive if you send away for information kits at up to $5.00 each, you will be saving more than money if you start with the airplane that is right for you.

Now let's take a look at some of the more popular airplanes that you can build from kits.

THE BD-1–BD-4

Jim Bede has just launched the most innovative homebuilt from his small factory near Cleveland. A single-seater powered by a small engine, it's test results are better than first calculated, which no one believed when they were published! Jim Bede was responsible for the original design of the Grumman American Yankee, which was designed as the BD-1 and in which new ideas were used to simplify construction and to improve performance while lowering the overall cost. Although the production model of the Yankee has, according to Bede, who left the company following a disagreement with stockholders over his management, compromised many of the BD-1's most desirable features, it has become one of the most interesting trainers and cross-country sport machines for the money. It is also approximately 40 percent faster than any other equivalently powered, general aviation aircraft.

The BD-2, another innovative design, was intended for an around-the-world trip. Because of a malfunction the trip was never made, but in the meantime Bede broke a number of existing records in the BD-2, including staying aloft for more than seventy hours without stopping or refueling. It is worth noting that the record-breaking BD-2 carries 565 gallons of fuel and is good for nonstop flight of 110 to 115 hours. To accomplish this, power is reduced by 10,000 feet, bringing the fuel consumption down to about 7.5 gallons per hour (gph). When the fuel burns off to the remaining 100 gallons, fuel consumption is down to 3 gph, using only 9 percent power, with 1,700 rpm and 7.5 inches of manifold pressure. Bede estimates that he can circumnavigate the world in just under 100 hours.

The starting point of the BD-4 design was a careful reevaluation of the BD-1 wing. That wing was extremely simple, consisting of a tubular spar (which doubles as a fuel tank), five ribs, two sections of wraparound metal skin, and a wing tip. It was bonded (space-age jargon for space-age glues that form a molecular bond) to obviate riveting, thus providing a clean surface; very little hand work was required for the wing's construction.

The BD-1 wing was economical, despite the tooling-up cost; light in weight; and produced high performance. The BD-4 wing is an improved version that features Bede's patented Panel-Ribs, which, according to his brochure for the airplane, are "stronger, weigh less, cost less, go together faster and easier" and in addition require no expensive tooling for their assembly. According to Bede, the wing is just as aerodynamically smooth, if not smoother, than the BD-1 wing.

The BD-4 was introduced in late 1968. Well over 10,000 people have purchased information kits ($4.50 each) and more than 2,500 have bought building plans ($40 each) as of this writing. Nearly 500 people have purchased materials from Bede Aircraft to start making their own BD-4s.

About 900 manhours are required to complete the BD-4, according to Bede, which can be powered with an engine ranging from 108 to 200 hp. With four seats, the 150-hp engine or larger is needed, giving a gross weight of 1,800 lbs (with the 150-hp engine) or 2,000 lbs (with the 180-hp or 200-hp engine). Useful load is best with the 180-hp engine at 920 pounds; cruise speed with this engine is better than 170 mph. Range at 75 percent power is 750 miles, with 45 minutes reserve.

The BD-4 has a compact appearance; a turbocharged version, N325BD, has the appearance of a very professional, high-speed touring machine for four. The interior is somewhat snug, but there is room for everyone in the plushly upholstered seats. One possible improvement would be to fit a see-through panel on the turtleback.

Construction of the BD-4 fuselage is of standard-size aluminum angle bar stock, bolted together with reinforcing gussets at predetermined locations. It is designed for rapid and easy construction with only the simplest of shop tools and doesn't require any aircraft fabricating experience to put together. After completion of the fuselage frame, aluminum skins (or fiberglass covering) are bonded to the aluminum angles with a neoprene-base adhesive. Pop rivets are used to secure corners in areas exposed to the airstream to prevent peeling.

The tail unit uses .025 aluminum, and the skin here is attached by normal riveting. All leading edges are preformed but require trimming and cutting to the final dimensions. The wing spar, by the way, is an extruded aluminum tube 6.5 inches in diameter, which gives considerable strength to the cabin and the wing panels. The patented Panel-Ribs mentioned earlier are modular fiberglass units and are supplied in semifinished form. The sections fit together so rapidly that two people can easily build a complete BD-4 wing in one day without any special jigs or fixtures. Individual sections can be sealed and turned into instant fuel tanks, as each panel rib holds eight gallons.

Bede recommends sealing six of the bays (three in each wing), providing a standard fuel capacity of forty-eight gallons. Actually, as many as ten of the panels can be used for fuel cells for the 150-hp engine or larger. However, if full fuel is carried, the cabin load will have to be reduced. On the other hand, Bede claims an incredible nine-hour range with full fuel capacity. As an optional extra, the BD-4 can be ordered with folding wings—this means it can be towed home and stored in the garage. Each wing takes about two minutes to fold; because of a safety device, it is impossible to lock the wing in position for flying without first making all aileron, flap, and other connections.

The original BD-4 prototype flew well, but the cabin was a bit uncomfortable for a person taller than five feet eleven inches. But the new ones have plenty of leg room and head clearance for two large men in the front seat, with good ground visibility. The eight-

inch nosewheel swivels fully, which makes the aircraft exceptionally easy to taxi, and the flexible aluminum landing gear works well
even on rough or muddy ground. You can also build the BD-4 as a
tail-dragger, i.e. two wheels up front and a tailwheel at the rear.

Pretakeoff cockpit check is standard, but the flap lever is
mounted above the pilot in the cabin roof—its unusual location
takes getting used to. Acceleration as the throttle is eased forward
is a pleasant surprise. Ground roll with the 180-hp engine is listed
as 600 feet, with clearance over a 50-foot obstacle listed as 850 feet
(at gross weight). With two persons and forty-five gallons of fuel
the BD-4 is much faster; rate of climb is almost 1,500 fpm. The
recommended climbout speed of 120 mph is reached very quickly
after leaving the ground.

At cruise speed, 24.5 inches of manifold pressure and 2,450 rpm,
an indicated reading of 170 mph is obtained. Twenty inches of
manifold pressure and 2,450 rpm give 153 mph. The controls have
a pleasantly balanced feel; the lightness of the ailerons and the
sensitivity of the flying tail make the BD-4 very easy to fly. For
fully coordinated turns a little rudder, rolling in and out, is needed.

Stall characteristics are about the same as on the BD-4's commercially produced competition. But a pilot who is not quick at
recognizing incipient stalls would be sensible to practice them for
half an hour or so. Correction is easy—forward stick and a touch of
power and you're flying again.

Noise level, as in most commercially produced airplanes, is too
high. But because this is a homebuilt, the noise level is largely
determined by what type and how much sound-damping insulation
is used. Ears are not self-repairing, and consistent exposure to high
levels of noise results in a slow and irreversible lessening of sensitivity of the eardrum. It is good practice to wear ear protectors,
which cost around $1.50 a pair and which with careful use will last
at least twelve months. The best ones are made of neoprene or
rubber and are preformed. Conversations with other people and
aircraft radio communications become much easier, although you'll
have to remember to speak louder.

Landing the BD-4 is easy, with very little pitch change when
adding the flaps. Visibility on approach—80 to 85 mph is good—is
excellent, and the flying tail makes for a smooth flare and gentle
touchdown almost every time.

Despite its critics—and there are several—the BD-4 is not a bad
airplane to look at and is easy to fly. As kit-built aircraft go, it
represents good value for the money and is a relatively inexpensive
and speedy transport. There are very few welded parts, which are

simple; any qualified welding shop can fabricate them. The Bede factory has plans to supply all welded parts finished. All in all, it's an easy airplane to put together—one builder fabricated the entire basic fuselage in three days! With a little help from friends a BD-4 could be flying in as little as six months. The BD-4 sells for $6,300 with the 150-hp engine and $7,200 with the 180-hp engine.

THE BD-5

With orders, which require a $200 deposit, numbering more than 3,000, the BD-5 has a customer commitment of more than $6 million, a pretty incredible state of affairs for an airplane that was not publicly demonstrated until the Reading Air Show in June 1973. This unique single-engine pusher monoplane is designed to fly at 200 mph from a 40-hp engine; it has a stall speed (flaps down) of only 48 mph. A higher-power version with a maximum sea-level speed in excess of 260 mph is planned, as is a jet-powered version that bears a price almost ten times that of the basic model ($2,100).

The BD-5J seems to be Jim Bede's equivalent of the Learjet. When Lear announced that he would develop and build a low-cost, high-performance bizjet, almost everyone in the aviation industry predicted calamity and woe. As each stage in the development of the Learjet was successfully completed, the experts confidently predicted that failure was just around the corner and that surely the next step would fail. They were even saying that people

FIG. 7.1

BD-5

wouldn't buy the airplane—if it was ever certificated. When some accidents did occur, the same experts blamed them on Lear. Poor design, they said, and poorer manufacturing.

Like Lear, Bede has gathered some of the most innovative thinkers in the business around him. But you'll hear the same bad-mouthing: "The engine doesn't work . . . it'll never fly . . . they'll just fall out of the skies in the hands of the average pilot . . . they ought to stop him from making it." Still, the BD-5 really does fly, and seems to fly well, and is slated to complete the very ambitious test program on schedule. More remarkably, not only does it appear that it will meet those projected design/speed parameters but may even exceed them.

The aircraft is a low-wing, single-seat monoplane; its fuselage is based on a high-performance sailplane. It looks very small when you stand beside it on the ground. In fact, Bede says the design was based on the same soaring aerodynamics used in the construction of the BD-2, basically a powered sailplane. In both aircraft the pilot is seated in a semiprone, rocket-type seat, as in European racing cars.

The entire design is unusual—some would say revolutionary. The engine is mounted flat on the bottom of the fuselage and faces aft. There's a fixed-pitch wood propeller driven by belts hooked to a variable-speed drive, giving increased performance for takeoff and climb. This does much the same job—at a fraction of the cost —of a variable-pitch or constant-speed propeller. The recommended variable-speed drive system is an optional extra and costs $90. Actual cost of operating the BD-5 works out to about 1.5 cents per mile, which is quite inexpensive for sport flying. The airplane is also stressed for aerobatic flight.

The BD-5B, which has a 21.5-foot wingspan, compared with only a 14.33-foot span on the 5A, is intended for soaring on days when modest thermals may be found. The maximum speed of the 5B with the small engine is only a bit slower than the 5A (204 mph vs. 212 mph) and offers approximately the same range of 1,000 miles with full fuel. The 5A is designed to ±12Gs ultimate and is limited to ±9Gs, which is a very strong machine. There is a reduction to +7.6 and −5Gs in the 5B. Both sets of wings can be ordered for an additional $150.

An electrical system weighing 9 pounds is optional ($145) and includes starter, battery, rectifier, wire, circuit-breaker switches, and terminals.

The BD-5 is a very modern flying machine. Its near-perfect aerodynamics are the result of lengthy study of every modern report on the subject and subsequent extensive analysis of engineering data, all of which was processed by computer; the BD-5 is the result of that final analysis. The airplane features 70 percent split flaps, wing spoilers, and tricycle landing gear. Spoilers give extremely precise control on approach and can also be used for extremely rapid descent or speed control.

Complete information packages about the BD-4 and BD-5 can be obtained ($4.50 each) from Bede Aircraft Inc., Newton Municipal Airport, Newton, Kansas 67114.*

THE COOT AMPHIBIAN

There are a number of homebuilders interested in a go-anywhere–type of aircraft; an amphibian is the nearest thing to it, besides a rotary-wing machine. The Coot Amphibian, designed by Molt Taylor, is a two-seat lightplane, with folding wings, that can be towed; it's not, strictly speaking, a kit, although the hull and foredeck can be purchased from Taylor.

Molt Taylor is a well-respected professional aeronautical engineer who is known for his work in producing a flying automobile, the Taylor Aerocar. The latest model looks and handles exactly like an expensive sports car. It is not, as he says, "just another airplane —or just another car. It's a whole new kind of transportation machine, and it's the only vehicle that can go from your home to my home at more than 50 miles an hour."

The design originated more than twenty years ago, and the Aerocar has been flown, improved, flown again, and improved again. The present version is known as the Model III. It comes in two parts: the sports car and a trailer, complete with wheels and hitch, which contains all the parts needed to turn the car into an aircraft. Taylor claims that it takes less time to put the two parts together "than it takes to carry your bags from your car to your airplane." Fail-safe interlocks are provided to each attach point; if they aren't fastened properly the engine won't start.

The same ingenuity that is apparent in the Aerocar is found in the Coot Amphibian. The cost of $3,500 is for the basic aircraft, without radio or instruments but including a secondhand engine and careful "scrounging" for parts. Price also includes certain "hard-

---

* *Air Facts* magazine has published an ongoing series of reports and letters about the BD-5 they are completing. 1973: May (p. 18), June (p. 27), October (p. 52). 1974: January (p. 4, 37), March (p. 6.), April (p. 94), May (p. 26).

FIG. 7.2

The Coot Amphibian

to-build" parts which are provided by the manufacturer, who also provides a completely comprehensive Bill of Materials with details of suppliers from whom the more difficult-to-find parts may be obtained.

Included in the overall price is a charge of $150 for drawings, materials list, and instructions amounting to more than 300 sheets that attempt to answer most of the queries that can arise. In addition, Taylor offers a free consultation service to any builder. If you think it will be helpful, you can also buy 100 4- × 5-inch professional quality construction photographs that show each stage of construction.

Exact performance figures vary with the type of engine used and the finished empty weight. But average figures are around 42 to 45 mph stalling speed with a cruise speed of between 100 to 120 mph. The airplane has a wingspan of thirty-six feet, its design concept deriving from amphibious assault gliders developed during World War II for the Marines. This float-wing configuration is extremely practical for rough-water operation, and the inboard portions of the wing panels operate as sponsons, providing adequate roll stability when the machine is at rest in the water. The Coot is a two-seat aircraft, but a seat for a small child may be installed in the baggage area.

You should have some basic manual training experience before you attempt to build this machine, but if you have it and think an amphibian might be a pleasant addition to the family, $3.50 sent to Molt Taylor, Box 1171, Longview, Washington 98632 will bring you three-view drawings, photographs, and an information packet. Construction time for the Coot Amphibian is around 1,500 hours; no special tools are required.

ARISTOCRAFT II

If you were looking for a new six-seat aircraft you'd expect to pay more than $30,000—most likely nearer $40,000. The Aristocraft II, which is offered in kits, costs just over $6,000, including a 180-hp Franklin engine. For families, airplane-campers, or hunters, this aircraft can mean the difference between just flying occasionally and having a real working tool.

The design is a development from the company's Model W "Winner," originated by the old Waco company in 1946. Terence O'Neill, president of the O'Neill Airplane Company, formed that corporation in December 1962, following the purchase of the original Waco prototype. He completely redesigned it, retaining and strengthening the basic structure. At a gross weight of 3,300 pounds, the Aristocraft II has a 6.9G yield design load structure, considerably in excess of many airliners; at 2,300 pounds the wing is stressed to 10G.

A number of different size engines may be used, ranging from 180 hp to 275 hp. With the latter engine you are permitted a gross weight of no less than 3,500 pounds. With the 215-hp engine, the Aristocraft II will cruise with six aboard at more than 130 mph.

If money is a problem, you can buy separate kits as you go along, only one of which costs more than $800. There are 575 square feet of detailed drawings, including full-size rib layouts, angle-indicator plans, material usage layouts (an extremely helpful assist to the newcomer) assembly instructions, and a detailed materials list. There are six different packages, which, including plans but excluding engine, total $3,335. If you buy them all at once you save $160. And if you buy the complete kit with engine, there's a further small saving.

The Aristocraft has been provisionally type certificated and flight tested by the FAA. If you need its carrying capacity, it could be the aircraft for you. Further information (for $4.50) can be obtained from the O'Neill Airplane Company, 791 Livingstone, Carlyle, Illinois 62231.

AEROSPORT RAIL

If you don't need to carry a lot of people a long way, don't want to do aerobatics, but would enjoy hopping around the patch once in a while, an airplane you might consider is the Aerosport Rail.

It's a single-seater, has reasonable performance, and is just about the simplest airplane to construct, without any sacrifice in standards of construction. But it is a bit unusual. First, it has two engines. Second, the cockpit is not enclosed—you sit on an exposed seat with the wind blowing across you, as if you were flying an early aeroplane or riding a motorcycle.

Gross weight is 700 pounds, and top speed at sea level is around 90 mph, but you'll probably prefer to settle for cruise speed, which is about 65 mph. Range is a mere 100 miles, so hopping around the peapatch is all you're going to be able to do. And with all that wind rushing around, you won't want to be up for more than an hour at a time.

Meanwhile, you'll have one of the sportiest of all light aircraft. Ground run on takeoff is a mere 230 feet, and the landing ground roll is 300 feet. So you won't be able to put it in anywhere you can't get out of! Stall speed is a little high at 45 mph but not unreasonable. Each engine comes complete with a built-in twelve-volt alternator, so you could fit strobe lights and other electrical equipment if you wanted.

The Aerosport company, which also makes a high-wing monoplane with enclosed cockpit—a very pretty bird called the Quail (to rhyme with Rail?)—offers a number of ways in which you can purchase either airplane. The complete package includes everything you need to build and fly your Rail, with the exception of gas and oil. Also included is a free tool package, airspeed indicator, altimeter, cylinder head temperature gauges, and tachometers. The company recommends at least twenty hours of solo flight experience for would-be builders, not because the airplane is difficult to fly but because it does take more concentration to fly an airplane without a cockpit than most beginners can muster. For those for whom the idea of open-air flying is anathema, the company has a new design on the drawing board with an enclosed cockpit— indeed, it may be in production by the time this book is published. For complete details of the Rail (or the Quail) send $4.50 to Aerosport Inc., Box 278, Holly Springs, N.C. 27540.

STURGEON AIR

Sturgeon Air Ltd., which purchased Falconar Aircraft Ltd. in early 1971, has become one of the major suppliers of homebuilding materials—steel tubing, plywood, sitka spruce, dopes, and fabric.

The company also handles the most comprehensive supply of kits for homebuilt airplanes in North America and very possibly anywhere in the world. Chances are that if you spot a homebuilt aircraft, at least some of the parts will have come from Sturgeon. And if you should see a two-thirds–scale Mustang fighter in your travels, that's also theirs, the world's first proven replica fighter. The response to the Mustang replica, according to the company, was so overwhelming that they decided to go ahead with research and development of a second scaled-down fighter. This is a three-quarters–scale Focke-Wulf FW-190A.

The Jodel series

Sturgeon also produces kits for a number of extremely successful basic types of homebuilts. If you have already had some contact with the homebuilding fraternity, you will almost certainly have heard about a machine called the Jodel, or of other machines described as Jodel derivatives.

Popular with sportsplane enthusiasts in Europe, especially in France and England, the Jodel D-11 series is found there in both factory- and amateur-built models. The basic airplane was originally designed using a 45-hp engine and was subsequently modified to use up to a 90-hp engine, with which it will cruise at 120 mph. If more power is desired—engines from 100 to 150 hp—one of the derivative models is recommended, such as the F11 or F12. But first let's look at the original Jodel.

The original Jodel was the "Baby Jodel," a snappy single-seat low-wing monoplane with a gross weight of less than 600 pounds. It was designed in France by engineers Joly and Delemontez (hence the name) and was already something of a success before World War II. It was designed to fly on as little as 25 hp and quickly became popular for low expense in construction, operation, and maintenance. In addition, it was a delight to fly and had no unpleasant or tricky characteristics. If it had a problem, it was simply that the experience couldn't be shared with anyone. Consequently, a two-seat, side-by-side version was designed, of which several hundreds have been built and flown in France, England, the United States, Canada, and elsewhere. It is one of the few lightplanes for homebuilt construction that is also in factory production.

Construction of the D-11 is simple and rugged, all wood, fabric-covered wing, elevator, and rudder; and light-plywood–covered fuselage and stab. The aircraft could be built directly from plans

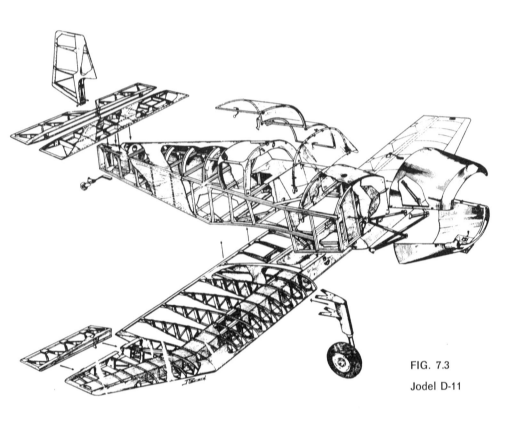

FIG. 7.3

Jodel D-11

($75), but kits are available too, for all or any part of the construction. In addition, in Canada the airplane may be licensed as an experimental aircraft or given a full Certificate of Airworthiness on subsequent special application. In the United States only the experimental rating is possible. The cost of a kit is around $2,500. (It's difficult to give a totally accurate figure, as there are so many items for a customized aircraft. The figure quoted is for a well-equipped aircraft of the D-11 series without engine. Range with the Continental C90-12 engine is 400 miles at 120 mph.)

Larger individuals should consider the possibility of the F-12, which has a slightly wider cockpit than the F-11. The F-12 is also a two-seat, side-by-side aircraft which has good range, is reasonably fast—cruise speed is 140 mph with a 140-hp Lycoming engine—and has a rate of climb in excess of 1,000 fpm at sea level. It is a rugged machine and can be operated from grass strips with no worries. It is available in a tailwheel version or with the more conventional trigear layout. The basic structure is birch plywood and sitka spruce, which are specially treated with a synthetic plastic finish before covering. This waterproofs and rotproofs the wood virtually for life.

Ease and simplicity of construction are key elements in the design, but it is a remarkably solid structure that is capable of withstanding as much as 9Gs. An important safety factor in the wing, for example, is that the most important highly stressed part of the spar consists of the four flanges stretching almost from one aileron to the other through the center section without a splice.

Flaps are optional, and for those who enjoy slipping, they may be unnecessary. But even without them, the stall speed is a low 41 mph. Aileron control is reported to be effective through the stall. Still, flaps permit steeper descents, and most builders add them. An extensive number of modifications are available in the form of supplementary plans, including swept fin and rudder, a sliding bubble canopy (which gives the airplane a modern, rakish air), auxiliary fuel tanks, and even a third seat—not yet legal in Canada. There's also a modification that permits the fitting of floats, and an optional three-piece folding wing can be fitted if parking's a problem.

Perhaps the only drawback is that the F-series tail and wingtips are slightly rounded and have an airfoil shape, as opposed to the Jodel's straight edges that have no airfoil at the tips. This does mean some extra work during construction, but in appearance the improvement is noticeable. The cost averages about $3,000, but with the Jodel and F-series, builders should work out their own individual requirements for a true cost—not merely in the performance parameters of their own individual aircraft but right down to such details as the material to be used to upholster the cockpit.

AMF-S14

The AMF-S14 is a high-wing, two-seat, side-by-side monoplane that with only 100 hp has truly STOL performance. If super-STOL is desired or needed, the airplane will take up to 150 hp, but most builders are more than delighted with around 125 hp, which gives a takeoff run of 200 feet or less, depending on the wind.

The AMF-S14 is especially popular with owners in the backcountry, as it is adaptable to wheels (tail or nosewheel), skiis, or floats. It has ample room for baggage or hunting or survival gear, and visibility from the cockpit is good. The folding-wing option can be useful, the more so since the floatplane version can be moored on the water with the wings folded with no fear of its being blown over.

Although it doesn't seem very large, the cockpit space is not bad at all. The noise problem can be cured by the careful installation of fiberglass insulation. In flight the aircraft appears very stable and is easy to maneuver. Cruise speed with the 150-hp Lycoming is around 120 mph and 110 mph with the 100-hp engine. Stall is a low 37 mph and occurs at a very high angle of attack. Slight buffeting in the model flown precedes the stall, and there was no apparent incipient wing drop. Aileron control is fully maintained through the stall.

The most expensive items in construction appear to be the engine and the propeller. The least costly AMF-S14 constructed to date was $2,000; expect a cost nearer to $2,850 using a kit.

Lobet Ganagobie

If money is really a problem, the Lobet Ganagobie may be the answer. This is a truly lightweight single-seater, which can be built for as little as $800, including the engine. It looks a bit ugly to the uninitiated, but is really very charming.

The Ganagobie has a gross weight of 460 pounds and will cruise between 50 and 100 mph, depending on the engine. Any engine in the power range of 20 to 72 hp can be used. Visibility is excellent, and control response is light and sensitive. The latter feature is unattractive—perhaps disconcerting is a better word—at first, as most of us are used to more sluggish control response. But the feel of the Ganagobie quickly becomes acceptable.

Designed and built by the brothers William and James Lobet from Lille, France, the first prototype flew in 1953. Several important modifications were made until an engine failure—they were using an antique Clerget, which weighed around 55 pounds—caused them to ground the aircraft, as no spare parts for the engine were to be found. It was then suggested that a surplus target drone two-cycle engine be used; with some reengineering they were able to continue their testing.

The next Ganagobie was a 72-hp McCulloch-engine version built by LaRue Smith of Delburne, Alberta, Canada. This had quite remarkably good performance, despite its being somewhat heavier, as it was built of birch plywood. This version is, incidentally, suitable for Volkswagen-engine conversions.

Perhaps the Lobet Ganagobie's most attractive feature is that it was designed as a low-cost, simple-to-build, ultralightweight machine for those who have neither the time nor the money to embark on anything more ambitious. It was also intended for the weekend

175

flyer who has no need to take friends or relatives at great speeds across the face of the earth. And although the performance can't truthfully be described as exciting, it is perfectly adequate.

Cruise figures vary according to powerplant, but with a twin-cylinder 35-hp Poinsard engine, rate of climb has been recorded at 350 fpm at an indicated 55 mph on a climb power of 2,200 rpm. Cruise speed on the same machine was 73-mph at approximately 2,150 and 2,175 rpm. You can't complain too much about that.

Construction is all wood, using sitka spruce and birch or mahogany plywood with a few pieces of hardwood. The four-sided, diamond-shaped fuselage is made up of thirteen easily shaped bulkheads and four principal longerons (longitudinal bracings). This structure is then plywood covered. You can have either square or rounded wing tips, and the set of plans shows the construction details of the optional swept-back fin and rudder, plus an arrangement for the installation and rigging of skiis. The wing spars are plain rectangular sections; the ribs are just a simple truss type. The wing is fabric covered.

If worried about the life of wood, don't forget that with the development of synthetic resin adhesives and fiberglass and plastic finishes, the use of wood as a construction material can be quite advantageous. After all, even metals are being glued together these days in aircraft construction! Wood is exceptionally easy to work with and doesn't cost too much. It has a good strength/weight ratio and when plastic coated will last almost forever. It also has good noise-dampening characteristics, provides good insulation, and will not corrode. Finally, it doesn't have the sort of fatigue problems common with many metal structures.

Detailed information on the Jodel D-9, D-11, F-11, and F-12, the AMF-S14, and the Lobet Ganagobie can be obtained from Sturgeon Air Ltd., 36 Airport Road, Industrial Airport, Edmonton, Alberta, Canada. Don't forget to send $4.00 for each model.

ROTARY-WING HOMEBUILTS

Three companies at the present time are marketing kits for home-built rotary-wing machines. The best known is the Bensen Aircraft Corp. whose Gyrocopter—a single-seat autogyro—is well known in this country and abroad. The firm also markets kits for gyro-gliders, which have to be towed over land or water, and they also market a float-equipped version of the motorized Gyrocopter.

The difference between an autogyro and a helicopter is that although each relies on its rotors to provide lift, the autogyro requires a propeller to provide thrust for forward flight, whereas the helicopter, using a cyclic pitch mechanism, changes the angle of attack of each rotor blade at a predetermined position to get forward, backward, or sideways flight.

If you can find some means of silencing the McCulloch engine on the Bensen gyrocopter, I would think it would be a most amusing toy to own. It has phenomenal performance, will take off in a mild wind in about 100 feet, and will almost land on a postage stamp. It is relatively cheap to build, and kits are available. Details can be had from Bensen Aircraft Corp, RDU Airport, P.O. Box 2746, Raleigh, N.C. 27602. Don't forget to enclose $3.50 for the information packet.

The two companies offering homebuilt helicopters are Helicom Inc., 4411 Calle De Carlos, who offer a one-seat machine, and Rotor Way Inc., 14805 S. Interstate 10, Tempe, Arizona 85281. Both charge $5.00 for their information package.

Unless I were an aeronautical engineer (and a helicopter pilot too) I think I would be rather wary about embarking on a pure helicopter design. It's not that it isn't possible to build an inexpensive two-seat helicopter—I'm sure it is—but I think one would need to know quite a bit about helicopter engineering before going ahead confidently. Although a helicopter offers incredible convenience in terms of its ability to land almost anywhere, its drawbacks are the very complexity required to produce this sort of performance.

In a helicopter the main rotor not merely provides lift but also provides thrust. This means that the blades have to change their angles of attack as they revolve. In addition, the torque that is generated by the relatively large main rotor has to be offset—and that usually means a second prop has to be linked to the engine.

HOMEBUILT SAILPLANES

Sturgeon Aircraft offers kits for homebuilt sailplanes, these being two models originally marketed by Falconar Aircraft Inc. One version may be fitted with auxiliary power. Glide ratios are 26:1 for the Fauvel AV-361 and 27:1 for the auxiliary-engine Fauvel AV-45. Both are flying-wing, single-seat aircraft.

The Schweizer company, whose name is almost synonymous with gliders in the United States, offers a kit version of their SGS 1-26 model, the SGS 1-26C at a price of $3,660. Gross weight is 575 pounds, and useful load is 220 pounds. The L/D (lift/drag) ratio for the 1-26C is 23:1, not startling but not too bad. Maximum speed is 104 mph. The 1-26 enjoys a good reputation and is a popular sailplane among those who prefer to rent rather than own.

8

Airplanes for Less than $5,000

The secondhand aircraft market offers good value in this inflationary age. Unlike used cars, airplanes are expected to be traded on in flyable condition. When buying a used airplane, bear a few pointers in mind. No matter what romanticists may say, be very wary of your heartstrings. Like a sweet melody, a neglected little airplane can tug at them, causing all sorts of emotions to well forth like some gushing mountain stream. Before you know it, you've given the owner the nod—and bought yourself a bag of nuts and bolts. It really is difficult to be sensible at times, but before you sign anything (let alone indicate that you think you might be interested) insist on permission for your own A&P mechanic to inspect the aircraft. Unless the owner has something to hide, he will agree. If a salesman is in charge and insists that his firm will guarantee the airplane, you could, without being rude, ask to see the guarantee. If that's not available, you are in your rights to have the aircraft inspected by your man. If the salesman says that company regulations don't permit it, don't waste any more time. Cross that one off your list, no matter how much you may want it. Let some sucker buy it.

Obviously, if you're buying a secondhand airplane, you won't expect the last word in modern design engineering. However, you'll still get something that will fly quite well, and at the price you should find it an economical first step in aircraft ownership; there's always the possibility that you may find a bargain.

Up-to-date logbooks in an airplane are one indication that the craft's previous owners have been well-intentioned and have attempted to cater to the mechanical vagaries of the machine you think you'd like to buy. Your general search for the sort of airplane you want (and you should decide what sort of craft you want before you begin looking) should not be limited. Classified advertisements in magazines, Sunday newspapers, even the weekly classified magazines—all these are fertile areas for search. The airplane you decide to buy may be right under your nose, but you'll only discover it by looking for it. Another good place to search is on notice boards in airport lobbies. Posting your own notices can be helpful too. And finally, as we suggested earlier, browsing around country airports sometimes turns up just the aircraft one desires.

All sellers seem to acquire rose-colored spectacles when they discuss their merchandise, and aircraft owners are no exception. The danger lies in accepting the seller's fantasies as fact. This is where your friendly A&P mechanic comes in. For an agreed sum (and don't ask him to do it out of friendship, or for a couple of beers—he has a right to earn a few dollars too) have him check the machine out. He will very quickly be able to assess what you'll have to spend to put it back into shape—the sort of shape you want. He'll also be able to tell you how much of this work you could do yourself, under his supervision. The airplane's records will give you a clue, but don't rely too much on them.

If the logbooks have been properly kept, you'll have details of all maintenance, any modifications, and whether the former owner has kept up with aircraft worthiness directives. Your A&P mechanic will be able to decipher the code and will probably know if all is as it should be. He will also be able to check on the alterations that may have been made but not recorded. And if you and he suspect that the records may have been altered, you can try to contact one of the former owners.

Documentation should include a current Certificate of Airworthiness, all aircraft and engine records, an equipment list, weight and balance data, maintenance manual, service letters, bulletins, airplane flight manual with details of operating limitations, and the bill of sale.

Title search is important; don't rely on the dealer to take care of this. The FAA can supply you with a list of title-search companies. The cost will be about $10, depending on the complexity of the case. If you belong to an association such as the NPA (National Pilots Association) or the AOPA (Aircraft Owners and Pilots Association), you can arrange to do this through them.

A good way to judge whether an airplane is worth checking out is to see whether its external appearance shows signs of care. Even if the paint is faded, a well cared for aircraft has a certain aura about it that is hard to define. Cleanliness is part of it, and a clean interior, though the carpet be worn, is but one indication that the owner is fond of his craft. Check inside the engine cowling. Cleanliness here usually means that everything is working properly.

Clues to tired airplanes are small holes drilled to prevent cracking of paneling or plexiglass. Other indicators are small patches on either fabric or aluminum. With fabric-covered airplanes—unless recently re-covered with one of the new synthetic fabrics now on the market—be careful. The cost of re-covering is sometimes more than the purchase price of the airplane itself.

Modern airplane engines are reliable. Still, caution should be exercised. A compression test on all cylinders will give some indication as to whether an expensive overhaul may be due sooner than you expect. Oil leakage or seepage can indicate improperly seated gaskets or more expensive damage, such as a cracked cylinder block or crankcase housing. Chipped or bent cooling fins on the cylinders or missing baffle plates are sometimes a clue that the engine is running hotter than the recommended temperature.

Replacement engines are not cheap, and even remanufactured engines are expensive. If there are indications in the engine log of repeated engine problems, you ought to add to your cost the possibility of some expensive repairs, if not an actual replacement. While you're checking the engine compartment, look at the inlet and exhaust manifolds for dents or cracks. Cracks in the exhaust manifold could lead to carbon monoxide poisoning. Discoloration of the manifolds could confirm that the engine has been running excessively hot.

Engine mounts may require replacing, especially if you notice vibration when you run up the engine. Mounting bolts should be secure, and a careful check of the wiring harnesses could save an expensive session with the electrician. Wires should be in good shape, free from fraying or wear or oil contamination. You should also check the fuel lines, especially at joints, to make sure there's no leakage and that the joints are properly seated. Wheels cost money; look to see whether the tires and the wheels they're attached to are in reasonable condition.

A thorough walkaround pays off—and if the airplane is a fabric-covered machine, have your mechanic bring a fabric punch gauge. Older airplanes used wooden propellers, and if your craft has one, check out the laminations and the leading edges. Compared with

181

metal, wooden props are soft—although propeller designers say you can make much more beautiful propellers out of wood than is possible with metal. If the propeller is a metal one, make a thorough check for any cracks, nicks, or dents. There's no reason why these shouldn't have been dressed, and if they haven't, it's a clue that the owner hasn't been too careful with his maintenance.

Walking outward to the wingtips, check the ailerons very carefully, especially hinges and the connecting rod. Look for rust and uneven wear. Your mechanic will check out control cables for slackness, wear, and security. If the airplane has tail-bracing wires, examine them to see that they are free from cracks or deformation and that they are securely fastened. Excessive play in the elevator and rudder hinges can spell trouble. All turnbuckles ought to be safetied with wire—if they aren't, you should ask why.

If the airplane is a tail-dragger, check the tailwheel carefully. It's mounted to a yoke and is sprung. Are the springs vibrant with life or are they soggy? And the tailwheel itself—how many more hours is it good for? If there's a locking mechanism for the tailwheel, check that it works properly; they have been known to go wrong.

You'll see now why it's simpler to take a mechanic with you. While airplanes are—to the initiated—almost as simple as cars, there are a number of unfamiliar items whose mechanical tolerances will generally be unknown to you. For this reason, go with someone who knows. Apart from anything else it will give you a chance to make a much more accurate estimate of the true cost of the machine you're buying. If you enjoy flying, you might as well have a machine that will serve you well.

Upholstering an old airplane is not such a difficult job, and there is a firm that can supply you with soundproofing material—all of which will contribute to the customizing of your particular aircraft. But make a very careful note of the weight of the materials you put in; although modern materials are very light, possibly even lighter than the original equipment, for your own comfort and safety you should know of any change in the weight and balance. As the owner of the airplane, you have a legal obligation to do so.

It's possible that you will want to update the instrument panel. Panels in older aircraft seem to have been assembled on a strange random principle. Nothing is in its logical position. The familiar "T" panel used in aircraft today works very well, although a number of engineers are trying to improve on that too. Unless the airplane you want to buy is a genuine antique, it seems to me that you're better off with as much information as you can afford; a panel that presents flight information in an orderly and easy-to-read way seems to have a lot going for it. Radios are discussed in Chapter 10.

So far we haven't touched on experimental aircraft. From time to time these are offered for sale, and generally they can be excellent purchases—if they fulfill your peculiar needs. But there are special rules regarding the purchase of experimental (or homebuilt) aircraft, as well as military airplanes. A talk with your nearest General Aviation District Office would be wise if you're contemplating such a purchase.

Before you may legally fly your newly purchased airplane, you will need a Certificate of Registration, FAA form 8050, which comes in three parts: Application for Registration, Bill of Sale, and Certificate of Aircraft Registration. Keep the yellow part of Part 1 with the aircraft and send the form to the Registration Branch, Federal Aviation Administration, Oklahoma City, Oklahoma, 73101. And now to some more aircraft . . .

Eclecticism may be applied to aircraft selection quite as reasonably as in philosophy, art, or science. Used airplanes in the market today offer the buyer an almost bewildering variety from which to choose and frequently a better value per dollar than might be found elsewhere. The following selection of aircraft does not pretend to be completely comprehensive and focuses on the more available aircraft.

THE TAYLORCRAFT

Soon after William T. Piper, Sr., bought control of Charles G. Taylor's company and the Taylor Cub, Taylor started up in business again, with a design similar to his famous original. A side-by-side, two-seater, high-wing monoplane, the Model B Taylorcraft was for a while faster than all its contemporaries except the Luscombe. Taylorcrafts are available today for a purchase price ranging from $1,000 to around $3,000. But they need careful checking before purchase, as all too frequently work is required to get them into flying shape.

Even experienced pilots are sometimes fooled into thinking the Taylorcraft is some funny sort of Cub, so striking is the resemblance. The original Model A, which was not replaced until 1939, was a near ringer, but its controls were infinitely superior to those of the Cub.

Perhaps one of the more endearing qualities of the Taylorcraft is that it is relatively simple to rebuild. Several now flying have had

FIG. 8.1

The Taylorcraft

their fabric replaced with synthetic material, but most are still in the rag and dope stages. If you decide to rebuild one, take care you check out the wooden stringers—these can require replacement. And while you're at it, check the spruce main spar—this sometimes needs replacing too.

If a secondhand Taylorcraft sounds hard to get airborne, take comfort that Univair Aircraft Corporation exists (Box 59, RR-3, Aurora, Colorado). Univair can supply just about any part used in the assembly of a Taylorcraft, right down to the original owner's manual ($6).

Although the original Model A had a 37-hp Continental, the Model B gave you a choice of a 50-hp Continental, Lycoming, or Franklin engine. In 1940 power was again increased, and in the last year before the war, with the addition of what were then called "luxury items" the aircraft was labeled Model BC-12. This version used a 65-hp Continental A65.

World War II brought a halt to civilian production, though the U.S. Army Air Force required no less than 1,000 aircraft from the factory for the duration. When hostilities ceased, the Taylorcraft reappeared with a strengthened fuselage (possibly a military by-product) and a modified number: BC-12D. In 1951 a speedier version, the Model 19 Sportsman, was offered. And there were also two four-seat versions on the theme, the Model 15 Tourist and the Model 20 Ranchwagon, both of which were introduced to compete with the other four-seaters of that time, such as the Stinson series.

One of the drawbacks to the Taylorcraft is that you must swing the propeller to get it started, unless an electric starter has been installed. In flight the BC-12 is docile and well mannered, although roll rate is a bit stolid in comparison to pitch response, which is very sensitive. Takeoff is at most a brief transition period into

184

flight, and there's sufficient rudder to keep this tail-dragger headed the right way in all but the strongest of crosswinds. The stall? Definite, but you won't get into trouble here. Cruise speed is anything from around 45 mph up to the magic 100-mph mark, and it will cruise hands off without much trouble. Noise can be a problem, but you could get ear-plugs if you don't want to attempt a soundproofing job.

The only real drawback is visibility—a problem you'll find with many airplanes of this era. Perhaps the best solution is to follow the pattern of the military version and fit plexiglass panels into the rear. For all that the Taylorcraft is a comfortable little airplane that will multiply your pleasures in flight in direct proportion to the care you give it on the ground. Versatile and economical, a Taylorcraft is hard to beat.

Specifications

Gross weight	1,150 lbs.
Empty weight	670 lbs.
Useful load	480 lbs.
Wingspan	36 ft.
Length	22 ft.
Height	6 ft. 8 in.
Fuel capacity	12 gals.
Range (cruise speed)	250—500 miles

Powerplant

Engine	Lycoming 0-145-B2, 65 hp
	Continental A65, 65 hp
	Franklin 4AC-176, 65 hp
Propeller	Laminated wood, fixed pitch

Performance

Takeoff	550 ft.
Rate of climb	450 to 600 fpm
Service ceiling	17,000 ft.
Maximum speed	105 mph
Cruise speed	95 mph
Landing speed	35 mph

CESSNA 120 AND 140 SERIES

Introduced in 1946, the Cessna 120 and 140 series of aircraft were surprisingly modern for their time. The controls are in almost the same place as in a recent 150, the principal difference being that the 150 has electric flaps and tricycle gear and the trim wheel is mounted by your right knee.

The 140 (See Figure 8.2) had more power than the 120, a starter-motor/generator kit, and manually operated plain-hinge flaps. Both were powered by the Continental C-85-12 and -12F engines; the 140A series came with the Continental C-90. The most unusual feature of the airplane—apart from the all-metal construction of all but the wing cover—was the use of tempered and tapered pieces of spring steel for the landing gear—the first time this was done on a production basis.

Inside you feel as if you were sitting in a sports car, as the seats are nearly level with the floor. But once your apprehension has disappeared, you'll discover that this really is quite a comfortable way to sit, and visibility over the nose is really quite good for a tail-dragger. The tailwheel steering is usually well balanced (if it isn't, have it checked out), neither too sensitive nor too slow. Tight corners require the use of brakes, but a touch of rudder is sufficient for most directional guidance on the ground. Performance is not outstanding, but it isn't all bad. Expect around 95 mph at cruise speed from the 120 and 115 mph for the 150 series, discussed elsewhere. The 120 and 140 will usually provide an honest 450-fpm climb at around 65 mph. Some owners of clean aircraft report better than 500 fpm.

Both the 120 and the 140 exhibit nicer control response than the 150. There's little if any lag and hardly any stall at all—some buffet, some mush, and a rather higher than normal rate of descent is all. Takeoff and departure stalls are not much more pronounced; visibility is quite good.

No airplane is completely perfect, alas, and the points to watch for in buying a 120 or 140 are corrosion and ground loop damage. The use of aluminum in the postwar years was a relatively new process for civil aviation at the two-seater level; standards were very low. The airframe itself, unless damaged or exposed to salt water or sea air, is virtually maintenance free. If it has been damaged, it is easy and safe to mend. The worst problem is corrosion—unless you want to rebuild, forget buying a corroded model.

Prices have remained relatively high for a design that's just short of the quarter-century mark. Anything from $2,000 up is being

FIG. 8.2

Cessna Model 140

asked, although with the recent upsurge in new-aircraft buying, secondhand prices should ease again.

Specifications

Gross weight	1,450 lbs.
Empty weight	770 lbs.
Useful load	515 lbs.
Wingspan	32 ft. 10 in.
Length	20 ft. 11.75 in.
Height	6 ft. 3.25 in.
Fuel capacity	25 gals.
Range (cruise speed)	450 miles

Powerplant

Engine	Continental C-85 and C-90
Propeller	Metal

Performance

Takeoff	680 ft. (minimum)
Rate of climb	500 fpm
Service ceiling	15,500 ft.
Maximum speed	120 mph
Cruise speed	102 mph
Stall (or landing) speed	41 mph

THE LUSCOMBE

Don Luscombe was an ambulance driver in World War I who was so taken with the idea of flight that he would trade cigarettes for a ride in aircraft being ferried to the front lines. In 1926 Luscombe became involved with the Monocoupe Corporation, which produced aircraft. By 1933 he had set up his own production company, which for the first time in history made use of die-cut metal construction, pioneering the method that makes our modern type of production with interchangeable parts a reality. The Luscombe Phantom Model 1, the aircraft born of these ideas, was an archetypical status symbol, with a reputation much like a Horsch or an Hispano-Suiza automobile of those days.

Finally certificated in early fall of 1934, the Phantom was equipped with electric stainless steel flaps, navigation and landing lights, an Eclipse starter, dynamo, and battery. In addition, the horizontal stabilizer was adjustable, controlled from the cockpit by rods and gears, and the vertical fin was cambered for directional trim. It was also theoretically fully aerobatic. There were a couple of problems, however. A price tag of $6,000 was one, and the narrow track of the cantilever mainwheels, which, combined with indifferent forward visibility on landing, contributed to a remarkable tendency for ground looping. The trouble was that the tailwheel would pivot on landing on a hard surface, and the cure—a steerable tailwheel—was not found. Twenty-five Phantoms were built, of which three are said to remain in existence.

A year before the Phantom went out of production, Luscombe introduced the Model 8—also years ahead of its time in design. Except for the wing covering, the Luscombe 8 (also known as the "Fifty," because its original powerplant had 50 hp) was of all-aluminum construction. A pretty airplane, the earlier models tend to be a bit cramped with their fixed-seat design, and both occupants will usually be glad to land for refueling before it is really necessary. Most of the early Luscombes around today are the 8A series, powered by the 65-hp Lycoming Model B engine, although a few prewar Models C and D with the 75-hp direct fuel-injection Continental may be found. (Fuel injection prior to World War II was almost unknown. The early Messerschmitts used fuel injection, but the first Spitfires didn't. However, the British soon got the message.) This latter model cruises at around 110 mph.

Production ceased in 1942, owing to priority restrictions on the use of metal for nonmilitary aircraft. The plant was enlarged and converted to the production of metal parts and assemblies for sev-

Specifications

Gross weight	1,310 lbs.
Empty weight	710 lbs.
Useful load	600 lbs.
Wingspan	35 ft.
Length	20 ft.
Height	5 ft. 10 in.
Fuel capacity	23 gals.
Range (cruise speed)	500 miles

Powerplant

Engine	75-hp Continental, 65-hp Continental, 65-hp Lycoming

Performance

Take off	650 ft.
Rate of climb	750 fpm (with 75 hp)
Service ceiling	15,000 ft.
Maximum speed	115 mph
Cruise speed	110 mph
Landing speed	42 mph

eral types of combat aircraft for the duration of the war. By 1946 the way was clear for the production of the 8E and 8F series, powered by the 85-hp and 90-hp Continental engine, respectively. A four-seat version, the Silvaire 11-A Sedan, was introduced two years later and continued in production (after 1954 by the Silvaire Aircraft Company) as the Silvaire 8F until 1960.

Any Luscombe is a pilot's airplane. A pilot checked out in one will have no trouble in flying a J-3 Cub, although the reverse is not necessarily true. There are numerous blind spots. The rudder doesn't have much feel, but if you're lined up at the numbers at the end of the runway you won't have any problems.

A Luscombe 8 series will run from around $1,500 for a tattered model 8A up to more than $3,250 for a Silvaire 8F in good condition. A good 8A shouldn't cost more than a thousand, an excellent value. Still, this is an aircraft that should be checked out closely.

189

In flight, control response is positive to light pressure. With the smaller engines cruise speed is down to 90 to 95 mph. The 65-hp version will just manage 400-fpm climb with two aboard, so the larger engine will give you an extra margin of security. All models glide pleasantly and slip well and will land perfectly at a steady 65 mph. Stalls? A positive break, but nothing to worry about.

An elegant airplane, almost pert in appearance, long-legged persons should try the seats for size.

THE ERCO, ERCOUPE, AND AIRCOUPE

Introduced as the Ercoupe Model 415-C in 1940, the Ercoupe was designed to be stall-proof and spin-proof. Fred W. Weick, chief engineer of the Engineering and Research Corporation (hence the original name), decided that the company's little two-seat, all-metal, low-wing monoplane would, by virtue of his design, be free from those characteristics that lead to the majority of air accidents —stall followed by spin. His design, accordingly, eliminated the necessity for rudder pedals (the rudder being connected to the ailerons), while providing insufficient up elevator travel to hold a stall. It's impossible in an Ercoupe to get aileron without activating the rudder automatically and by the right amount. In fact, flying the Ercoupe is rather like driving a flying car.

The airplane has been produced under a number of names, including the Forney, Alon A-2, Mooney, and Aerostar Cadet. The last two received a redesign job enabling them to both stall and spin, as it was felt that the Ercoupe could also make a good trainer. It seems probable that it will be resurrected in the near future, possibly with a new engine of the rotary variety.

The Ercoupe is a pleasant little airplane. The cockpit canopy can be opened in flight, and the absence of the rudder pedals—once you get used to it—is not unbearable. If you really want rudder pedals, they can be put in. Because the earliest of the breed—the Erco 310, with a 55-hp engine—flew in 1937, the numbers of modifications that have been made are legion. Mooney, for example, replaced the twin tail with a conventional stern and made major changes in the landing gear. The earliest models had fabric-covered wings, but a number of owners have since had these re-covered with metal.

If the Ercoupe does have a fault, it's the high sink rate, which requires full forward yoke and plenty of altitude to pick up. Otherwise its basic design is such that it'll cover the mistakes of the most ham-fisted pilots. You can expect to pay around $2,000 for a postwar 75-hp model, less for the prewar model. The newer 90-hp Alon and Mooney Ercoupes cost more; expect anything up to $5,000. The average price—including the Forneys—was just over $2,000 at the time of writing.

Although it is only a two-seater, the Ercoupe was always faster than its contemporaries and will buzz along very pleasantly all day at 100 mph, at about twenty-five miles per gallon. It handles crosswinds nicely with it's castoring gear; about the worst thing you can say about an Ercoupe is that it was years ahead of its time.

Specifications

Gross weight	1,260 lbs.
Empty weight	725 lbs.
Useful load	535 lbs.
Wingspan	30 ft.
Length	20 ft. 9 in.
Height	5 ft. 11 in.
Fuel capacity	23 gals.
Range (cruise speed)	525 miles

Powerplant

Engine	Continental A65, 65 hp
Propeller	Fixed pitch laminated (late models: aluminum)

Performance

Takeoff	900 ft.
Rate of climb	500 fpm
Service ceiling	13,000 ft.
Maximum speed	115 mph
Cruise speed	105 mph
Minimum speed (power on)	42 mph

STINSON VOYAGER AND STATION WAGON

In the summer of 1940 the Stinson Aircraft Division of the Aviation Manufacturing Corporation was taken over by Consolidated Vultee; the principal World War II products were military variants of prewar civil airplanes—the L-5 Sentinel, whose predecessor was the prewar Stinson Voyager, and the AT-19 Reliant. The latter was a big five-seater that enjoyed use among businessmen and sportsmen who couldn't afford Beech's Staggerwing.

Both aircraft continued in production once the war was over, but the Stinson 108 Voyager and the Station Wagon were, for a brief period during which more than 2,000 were sold, the most sought-after of light aircraft. There are basically four models in the 108 series, and the principal differences are either in powerplant or rudder. Most of the 108 and 108-1 series make use of the 150-hp Franklin engine; the 108-2 and -3 models make use of the 165-hp Franklin. The Stinson 76 Sentinel, or L-5, as the Voyager was designated by the military, used a 190-hp Lycoming 0-435-1 six-cylinder engine, with a cruise speed of about 130 mph. Two wartime variants were made—the first was for short-range liaison and observation, and the second for ambulance work and light cargo. The 108-3 models are more easily distinguished, as the vertical fin was generously increased to provide added directional stability. This turned out to be a mixed blessing: The big fin acts like a sail in heavy winds if you're not careful.

A majority of the 108 series seems to have been re-covered with metal of fine-gauge aluminum, which does away with the bother of having to refabric every seven years or so. Several owners go much further, completely updating the interior—which was always a bit dark—and catering to creature comforts in new and adjustable seats. If you should purchase this type of Stinson, you'll pay around $7,500 for what is virtually a new airplane. The cheaper and older ones run about $3,250 and more. But when you are checking out Stinsons, be wary—the dealer at the other end could be wasting your time.

For a going-places vehicle, it offers an honest 120-mph cruise speed for four people, with a range of around 500 miles. Voyager and Station Wagon seem somehow very appropriate names for these old-timers.

Specifications

Gross weight	2,400 lbs.
Empty weight	1,294 lbs.
Useful load	1,106 lbs.
Wingspan	33 ft. 11 in.
Length	25 ft. 2 in.
Height	7 ft. 6 in.
Fuel capacity	50 gals.
Range (cruise speed)	554 miles

Powerplant

Engine	Franklin 6A4-150-B3, 165 hp
Propeller	Metal

Performance

Rate of climb	800 fpm
Service ceiling	14,000 ft.
Maximum speed	146 mph
Cruise speed	130 mph
Landing speed	51 mph

GLOBE SWIFT

The Globe Aircraft Corporation was originally formed as the Bennett Aircraft Corporation and intended to manufacture aircraft using a material called Duraloid®, a phenol-formaldehyde bakelite-bonded plywood. Reorganization in 1941 brought about the change of name, and the original Swift (Model GC-1)—a two-seat, low-wing cabin monoplane—received an Approved Type Certificate in the spring of 1942. The war stopped plans for production, and the Globe Aircraft Corporation turned out some 600 Beechcraft AT-10 twin-engine trainers and did some contract work on C-46s before resuming production of its own.

The Swift is a neat airplane, capable of a cruise speed of between 120 and 150 mph, depending on the engine. One owner has a supercharged Swift that sends Comanches and Mooneys scurrying out of the sky in terror (this Swift is supposed to do a neat 230 mph at cruise speed). It's aerobatic and is stressed to +7.35 and −3 Gs.

The only maneuver you want to watch is the snap roll, as a low entry speed is important. The tail is the weakest point, and too many snaps could be too much for it.

The retractable-gear motor has been known to cause problems; it appears to be too weak for the task. Three-point landings can also bring tears to your eyes—though it willl wheel land with no problems at all. The trick for a three-pointer is to keep your hand on the flap selector and spill the whole cushion of air just at touchdown. As for left crosswinds—the rudder, we must be honest, is just too small to handle these on takeoff.

The Swift can be bought for as little as $3,250 to as much as $8,000 for one in fine shape. Swifts are inexpensive to fly, about $5.50 an hour for gas and oil. You can add auxiliary tanks to extend your cruise range up to five hours. There's not much room; you might say that the Swift is basically transport—and fun—for one person.

Stalls are quite normal, but spins are prohibited. If you're looking for a fast, point-to-point fun machine, a Swift may be the one.

Specifications

Gross weight	1,710 lbs.
Empty weight	1,139 lbs.
Useful load	580 lbs.
Wingspan	29 ft. 4 in.
Length	20 ft. 10.75 in.
Height	5 ft. 10.5 in.
Fuel capacity	30 gals.
Range (cruise speed)	420 miles

Powerplant

Engine	Continental C125, 125 hp
Propeller	Metal

Performance

Takeoff	700 ft.
Rate of climb	1,000 ft.
Service ceiling	16,000 ft.
Maximum speed	150 mph
Cruise speed	140 mph
Minimum speed (stall, landing)	57 mph

THE NAVION

The end of World War II saw the resumption of light-aircraft production. Although many of the aircraft had been in production before the war, the Navion was a dully new design, the product of a company that had avoided the light aircraft market before.

The Navion is an all-metal four-seater, a low-wing monoplane with retractable tricycle gear. A well-used Navion will still give you a true 160-mph cruise speed; when new, the 285-hp engine gave something on the order of 190 mph. It's comfortable and quiet (even by today's standards), it's a good, stolid instrument platform, robust and easy to fly. Its mien is a bit stodgy, but there's a reassuring business-like air about its overall appearance.

Like many postwar designs, the Navion's changed hands. Originally built by North American (of Mustang-fighter and Mitchell-bomber fame), it was then sold to Ryan Aeronautical, who sold it

Specifications

Gross weight	2,750 lbs.
Empty weight	1,782 lbs.
Useful load	1,315 lbs.
Wingspan	33 ft. 4.5 in.
Length	27 ft. 3 in.
Height	8 ft. 7.5 in.
Fuel capacity	up to 7 hours range
Range (cruise speed)	500 miles

Powerplant

Engine	Continental E-185-3, 205 hp, Continental 10-470, 260 hp Continental 10-520B, 285 hp
Propeller	Metal

Performance

Takeoff	750-1,700 ft.
Rate of climb	900 ft.
Service ceiling	15,600 ft.
Maximum speed	163 mph
Cruise speed	155 mph
Minimum speed (stall, landing)	55 mph

to Temco, who in turn sold it to the Navion Aircraft Company. The revised version was called a Rangemaster, and with tip tanks had a flight endurance of about seven hours. (An excellent book called the *Navion Buyers Guide* can give you all the ins and outs of the Navion story. It costs $2.50 from the American Navion Society, Box 1175, Banning, California 92220.)

From the point of view of comfort, the Navion is spacious and visibility is excellent. The controls are effective, if a bit on the heavy side; very little rudder is needed to coordinate turns. Stalls are a mush, rather than anything more deliberate. Not surprisingly, Navions are rather hard to come by. You can expect to pay $5,000 or more. As a luxury touring craft with a range of more than 1,200 miles, the Navion has a lot going for itself.

PART V

9

The Real Costs of Flying

A lot of people have the impression that flying is very costly and is only for the very rich—and judging by the equipment the FAA would mandate that every aircraft carry, it could get to be like that one day. But at the present time the costs of flying compare very favorably indeed with the costs of other forms of recreation and sports, or of transportation.

A four-seat retractable can cost less than some of those fancy motor homes you see rolling down the highway, although you won't have room for a closet and the kitchen sink in the airplane. But there are several potential tax breaks for the aircraft owner, and your accountant can advise you on which would be best for you. Insurance compares favorably with automobile insurance, and real depreciation on aircraft is rather similar to that on Rolls-Royces and fine sailing yachts. It exists, and it tends to work in your favor.

One way to estimate the cost of your flying would be to check with your local FBO and see whether he'd provide you with a package rate (rather like renting a limousine) if you agree to buy so many hours per year. Now take that rate, and see whether the average annual number of hours you've flown over the past three years makes this more or less economical than your actually owning a similar type of aircraft.

Before getting into detailed analysis of whether ownership is cheaper, it may be helpful to see why it is that airplanes have to cost so much initially. The following is the breakdown of a $21,000 airplane.

Raw materials (airframe, appointments)	$ 1,250
Systems, instruments, hardware (wheels, braking systems, cables, bellcranks, flap motor, bearings, propeller, etc.)	3,000
Cost of engine to manufacturers	3,500
Subtotal	7,750
Labor—airframe	2,000
Labor—other	600
Subtotal	10,350
Manufacturer's overhead, general and administrative costs, taxes other than Federal tax on profits.	4,250
Manufacturer's markup.	1,360
Dealer's markup	5,000
Total	$20,960

You will note that the reason a $21,000 airplane is as expensive as it is is that a lot of people are earning money on top of the real cost. Before the airplane has left the manufacturer's hands it has doubled in price, by reason of the heavy burden the manufacturer must impose on it, in addition to his relatively small profit. The markup for the dealer appears high, but in fact the trade rule-of-thumb is that a minimum of 15 percent profit is needed to cover overhead, and the remainder—10 percent—is dealer profit.

One area in which complaint might be made is in the high cost of engines and of ordinary automotive parts when they are supplied to airframe manufacturers. The reason for the cost of engines, all of which, including old designs, have been rising steadily, is the need for engine-makers to cover their own overhead and research, as well as a fairly hefty middleman's commission.

An airplane engine must run without fault at near peak power for most of its life, which is anything from 1,400 to 2,000 hours before overhaul for a normal piston engine. If we assume a modest cruise of around 150 mph, the 2,000-hour engine will have run no less than 300,000 miles before requiring some serious attention. (Jet engines have even longer lives and consequently cover even greater distances between overhauls.)

Unlike the car engine, the airplane engine must be light and compact. It has also to be air-cooled and is not supposed to overheat or overcool. It's rather as if you were expecting all automobile engines to perform like the fabled Japanese cars. But airplane engines are putting out this sort of reliability day after day, and the manufacturers consequently expect you to pay rather more for such performance.

If an airplane is treated to an ongoing program of regular maintenance, most will last for fifty years or so before you have to get rid of them. There are still large numbers of aircraft flying that are more than thirty years old—how many automobiles that age have you seen around recently? And this is a point that is often overlooked by many would-be owners—if you don't think you can use a brand new airplane a sufficient number of hours per year to make it economical and you can't share with sufficient people to make it viable, you can still afford to be an airplane owner and enjoy cheap flying with a secondhand airplane.

The normal dividing point between renting and owning an airplane is 250 to 300 hours of annual use. A used airplane, since the actual investment is often considerably less, enjoys a lower overall cost structure since depreciation is not such a high factor. It can be more economical to own a small, used airplane to fly little more than 130 hours a year than to rent a new four-seat model for the same number of hours.

DIRECT AND FIXED COSTS

At this point there are two types of calculation you should make in estimating your costs. The first will deal with the actual running cost of the airplane, the second with the indirect cost of ownership.

Figuring out fuel consumption in an aircraft is relatively easy; you can work out before you start what you're likely to use. In your calculation you'll figure to burn so much on the ground from the ramp via the taxiways to takeoff. Then you can figure another amount from takeoff to cruise altitude. Then figure for the cruise portion of your journey and for the landing—and for an additional forty-five minutes of flight for spare. You must also calculate oil used, including oil changes every twenty-five hours. (Some manufacturers recommend oil changes every fifty hours, and which seems to work well with some of the fine oils now available.) Your cost will vary from a low of around $2.00 to around $5.50 an hour for a single-engine retractable. (A Mooney Mite, on the other hand, can spoil those figures, operating at considerably less than a penny a mile while averaging some 35 mpg—at more than 120 mph. Unfortunately, it's only a single-seater.)

Apart from actual fuel and oil consumed per hour, you will want to budget for normal maintenance, which will include an annual inspection if you don't rent the airplane to others or an inspection every 100 hours if you do. Maintenance will run from about $1.60 to about $2.75 an hour, and you should add in an extra dollar every time you change the oil. It is much better to check with your friendly A&P and see whether you can organize a progressive maintenance program rather than merely complying with the letter of the law. Under progressive maintenance your airplane is almost always in top condition, and there is a much better chance of catching small problems at the earliest moment. And there is nothing more foolish than flying an airplane that has things wrong with it. Even if it's only a faulty gauge—small malfunctions can get out of hand.

In addition, you should put aside a reserve fund to cover the eventual replacement of the engine, propeller, and perhaps even the tires. This figure is easy enough to calculate if you've bought a new airplane; you simply work out what the Time Before Overhaul (TBO) figure for the engine is and divide it into the replacement cost. With a used airplane you may need more. As for tires, how much you'll need to charge yourself will depend on your own take-off and landing technique—and on the runway surfacing material at your main base. Some rental operators expect to replace tires every 300 hours. Some private owners manage much better.

The costs outlined above are direct costs; they go up as you fly more. Indirect costs are fixed; these go down the more you fly. Fixed costs include such items as hangarage, or tie-down rent, and insurance for hull and liability. And strictly speaking, your annual inspection—which is required for the issuance of the Certificate of Airworthiness—should also be listed here, as you need it even if you hadn't flown the aircraft the preceding year. Finally, there is depreciation and the amount you have to pay back to the bank, if you borrowed from one.

These indirect or fixed costs are divided into the number of hours you actually fly each year. You then add this result to the hourly amount from the direct-cost figure to find out your total cost per hour. Let's see how they might add up.

If you have an arrangement with the bank, you will have to carry hull insurance. The bank (or finance company) will insist on its being carried to protect their part of the investment. But many owners of old aircraft don't bother with hull insurance, since the older the airplane, the more expensive hull insurance is. Depending on your own experience and the newness and cost of your airplane, you can expect to purchase hull insurance from as little as 1.5 percent of the hull value to about 5 percent. The first figure might apply if you were an ATR pilot with about 10,000 hours and your airplane was a 1972 model of the Cherokee Six. But were you trying to insure your Stinson Station Wagon, and had only 300 hours and a private ticket, you would be fortunate to find it costing you only 3.5 percent of the hull value.

It is well worth the trouble to find a broker or an insurance company that specializes in aviation insurance rather than buying from your usual company. You willl almost certainly get a better deal with a firm that has experience in the field, so you should shop around. Quite often one firm may have better experience with one company's particular model than another and will offer you a lower rate. Hull damage, by the way, covers any damage sustained on the ground or in the air and is usually quoted as "All Risk".

Liability insurance is also needed. It makes good sense to buy all you can afford. Liability insurance is usually based on the number of seats in an airplane. Single-liability policies for $1 million are not expensive for a four-seat airplane, and if you can find it, a $3-million policy (the highest quoted for general aviation light aircraft at present) does not cost that much more. (For the aforementioned Cherokee Six you'd expect to pay between $300 and $400 for a policy; a policy for a two-seat aircraft would cost perhaps half that figure.) With liability coverage if your engine should quit in the air, you don't have to worry about hitting a dozen chickens if you have to use farmer Giles' free range for your landing. It is also possible to buy the $1-million, single-limit coverage for yourself for only $50 a year, even if you don't own an airplane. (Some airplane renters will give you a discount for this sort of coverage.)

The price of a hangar varies enormously. A tie-down near an urban area can cost half as much as garaging a car in Manhattan by the month. Elsewhere you can get a proper hangar for that price, and maybe even less. If you own your airstrip, you can erect a hangar for a modest sum; T-hangars are quite inexpensive. Smith & Wesson, the famous gunsmiths, produce an excellent fiberglass T-hangar for around $1,500. How to include this in your cost estimate is up to you; perhaps the most convenient way would be to write it off over seven years and add the per annum charge into your fixed costs.

The convenience factor of private air transport and the relative savings compared with traveling by car are important considerations. For an executive earning $25,000 a year for a 2,000-hour working year, the relative cost could work out as follows. This person is being paid at the rate of $12.50 an hour; in the course of his job he must travel, let us say, some 35,000 miles a year. In a car at an average speed of 50-mph this represents some 700 hours of this person's time, or $8,750. Now let us put this executive into an airplane. It needn't be very fast. Let's say a four-seat, single-engine machine with a cruise speed of 130 mph. The executive can now cover that 35,000 miles in a mere 269 hours, for a saving of almost

$5,500. In fact, if the additional seats in the airplane can be filled some of the time, the savings will be even more.

There's an additional factor here too. Many times the executive with an airplane will not have to stay overnight at a motel—added expense—but can return home following the meeting. So from the business point of view, aircraft can make good sense. We won't go into a comparison with the airlines—it isn't really necessary, as their frequent-service schedules are confined to about twenty-two cities across the nation.

RENTING

Renting an airplane can be convenient. Perhaps the only drawback as far as low-time pilots are concerned, is the renter's insurance policy. Usually this specifies a minimum of forty to fifty hours experience with retractables, and a minimum of 250 to 300 hours experience before a person may rent. Lease-A-Plane International (LAPI) which seems to add new hubs to its operation as each year goes by, does not require a specific minimum total time from pilots wanting to rent fixed-gear aircraft. To rent a retractable, although no set requirement for previous retractable experience is mandated, a minimum of 125 hours total time is. As is a five-hours-in-type, plus a thorough checkout. For twin-engine aircraft a minimum of 500 hours total flying time is required. From the point of view of renting, LAPI airplanes are always pretty new, being traded well before major overhaul time—depending on type, between 12 and 18 months on average. Equipment is moving towards standardization of dual transceivers, ADF, transponder, marker beacon, and glide slope. Some aircraft have autopilots too. One further advantage is that when you are checked out in a type, LAPI issues you with an identity card that states which type you've been checked out in. Provided you stay current and don't mind sticking to that type of airplane, you won't need a checkride at other LAPI hubs.

Renting is useful for the person who doesn't get to fly too many hours each year; its only drawback is the necessity to keep current. Quite frequently, if you are only concerned about transporting people from point A to point B, renting a six-place airplane can work out cheaper than traveling by commercial airline. True, you may not travel there quite so quickly, but you may discover that the door to door time is not so much more than if you flew commercial.

The person renting an airplane will usually be asked to pay for a minimum number of hours per day—depending on how long a period the airplane will be away. LAPI operates a different sort of scheme, more like a car-rental firm. It may be $20 per day and 20 cents a mile (with a minimum of 450 miles, say, per day) or a flat rate of $60 per day plus $10 an hour. When you get back, they work out whichever way is cheapest. And you don't have to pay for time on the ground because of weather. Finally, you should be able to provide an adequate credit history or you may be asked for a deposit.

Renting charges also vary considerably according to the renter's overhead. In rural areas you may be able to rent a Cessna 150 for as little as $10 an hour, whereas near an urban area you may be asked for nearly twice that amount. If you rent regularly, you may be able to work out a deal for a discount—for example, agreeing to fly a minimum of ten hours each month. Or you might be able to arrange to get ten additional hours in an agreement to buy and pay for 100 hours. Once again, if you want to rent, shop around. Most small FBOs can offer quite good rates, and their aircraft are usually very well maintained, as they frequently use them themselves.

LEASEBACK

Another means of acquiring an airplane is by leaseback. Leasing arrangements are very popular with the airlines; by leasing aircraft they do not tie up their capital in equipment. There are also tax advantages to the owner of the airplane. So far most leasebacks seem to be in the western states, especially California.

Basically, leaseback is a contract with an FBO in which you agree—in return for a cash consideration, as the lawyers put it—to supply his firm with a rental airplane. Obviously, the FBO is going to try to rent the aircraft out for as many hours as possible, so your bird will log a lot of hours. On the other hand, by leasing the airplane to another, your indirect costs are very substantially reduced by reason of your rental income.

Although you might expect to save from one third to one-half of what it would cost you actually to rent an airplane elsewhere, the biggest benefit in leasing out your airplane comes in terms of your income tax. At the present time the law says that a lessor of an airplane is in the business of leasing aircraft, even if he only has one airplane. This means that you can immediately write off your depreciation on the airplane (which you would not have neces-sarily been able to do as an ordinary owner), as well as other costs. As an extra bonus, you are permitted to use the Investment Credit provision of the law. This means that you may actually subtract from the tax you would owe, even after you'd made your per-missable deductions.

Now before you go rushing out to buy that new airplane, let's take a closer look at leaseback. First of all, forget it unless you are already paying at least $2,000 or more in tax. Next, you must be sure that you'll be doing business with an honest person. Third, you'll want to talk with your tax accountant—you ought to get professional advice on what type of arrangement you need that can best suit you.

The best sort of deal to look for are those in which you get a decent discount on the purchase price of the airplane and the equipment that you'll put into it. You should also have some sort of provision to limit the amount that can be spent on repairs and maintenance without your approval. Once again, this is an area in which a preventive maintenance program combined with progres-sive maintenance can pay off. The aircraft should have a daily squawk sheet (completed each time it is flown) so that such items that do occur can be checked each time the airplane goes into the shop.

In terms of compensation there are a number of methods to choose from. You may be guaranteed a minimum number of hours usage per year—whether or not the FBO can meet them. Or it may be a straight one or two-year lease. Whatever the arrangement, insist that all the books and the accounting be available for your— or your nominee's—inspection at any time.

Because the aircraft will be building up hours at a pretty fast rate, the question of major overhaul assumes greater importance. Those experienced in leaseback suggest that there are two or three ways of dealing with this. You can go ahead, cost in the major overhaul and continue leasing the machine. Because there will be newer airplanes around by the time this happens, you may have to cut back on your price in order to maintain the same number of hours usage; indeed, even with that you may suffer some reduction in hours rented. You can major it and then sell it or turn it in against a new airplane. Or, before the aircraft comes up for major, perhaps at 1,450 hours on an 1,800-hour TBO engine, you could trade it in against a new airplane.

For most of us, leaseback involves an airplane like a Cherokee Flite Liner or Cessna 150 at the lower end of the scale; a Grumman

American Traveler, the Cessna 172, or the Cherokee Challenger in the middle range; and an upper level of the larger (and more expensive) fixed-gear machines. Although opportunities exist for retractables, for the very reasons cited earlier (the insurance companies requirements on flying experience before you can rent them) the rental hours tend to be less. Obviously, if your tax situation can afford it, you could arrange a leaseback arrangement with a Learjet—as a number of corporations do. While not so useful for the individual, the tax advantages might conceivably be such as to make it worthwhile, and you'd have the advantage of using it yourself for rather less than you'd be able to rent it elsewhere!

The other important question is the equipment you're going to install in the aircraft. If you want to rent the aircraft as much as possible—since the FBO is going to be able to make money that way too—you naturally will want to equip it as well as you can. This means that adjustable seats, for example, are worth the additional cost. Similarly, you won't want to skimp on such items as a good instrument panel, decent lighting, or True Airspeed corrector rings. Naturally you'll put in good radios and avionics.

The sort of minimum you might fit in a Cessna 172, for example, would be a 720 navcom with localizer and glide slope, three-marker beacon, and transponder. You might possibly add an ADF or a second navcom, say a 360 channel. Similarly, you should make sure that the overall instrumentation was set up for IFR work, and perhaps add such options as long-range fuel tanks, which can add to the utility of the aircraft. Or you could purchase the Cessna Skyhawk II, in which IFR capability is provided already at a large saving in the purchase price.

Your actual outlay for such an arrangement need not be excessive. You could reasonably expect to finance 80 percent of the purchase price of the aircraft for a period of five years. Interest rates vary, but you should expect to add on a further 5 or 5.5 percent of the total amount borrowed for the period of the loan. (For example, if you financed $10,500 on the purchase price of the airplane, your total repayment would amount to slightly more than $13,000, at five percent. This is not a great deal for the use of someone else's money and works out at a repayment program of less than $220 per month for the period.)

We're talking now in terms of a training aircraft, something like the Cessna 150, the Piper Flite Liner, or the Grumman American Trainer. Figures for the Beechcraft Sport, which is slightly more expensive, would be only a little higher. To talk in round figures, let us assume that our repayment on the aircraft is going to cost about $220 each month and that our insurance (hull—all risk plus $1 million single-limit liability) comes to $75 a month. We must add the parking fee (although we may not be obliged to pay for this), which we'll say costs $25 a month. This gives us a total indirect cost of $320 per month, to which we may be inclined to add a 10-percent contingency allowance, bringing it to $352—say $350 per month.

NEW AIRPLANE PRICES

Now let's take a look at new airplane prices. A Grumman American Trainer costs about $9,500 basic. A Cessna 150 Commuter (which can double as a trainer in its spare time) costs about $12,500—the basic 150 is about $1,200 less. Piper's Flite Liner costs $15,000, and the Beechcraft Sport tips the scales at around $16,000. Let us say that we shall want to spend a further $1,750 on radios. If we now average these figures we are talking in terms of $13,000 plus, to which we add radio equipment, giving us in round figures a potential investment of $15,000. Let us assume that we're in a position to get a discount to bring the figure back to between $13,000 and $13,500 as our real purchase price, of which we've been able to finance $10,500. So far, then, our total real outlay has been around $3,000, plus whatever local taxes we may have had to pay.

One last word about financing—it really is worth shopping around for the right arrangement. Like insurance, if you are prepared to put some time into finding a better deal, you will almost surely be rewarded. Firms that specialize in aircraft financing will usually be able to give you a better deal than those that don't, but occasionally if you have an understanding bank manager, you may actually get a better deal from your own home-town bank than from any of the bigger firms.

An airplane that is earning its keep (that is, is being flown for money) must have a regular 100-hour inspection that is signed off by an A&P mechanic, unless it is being maintained under a progressive maintenance program. For gas and oil we might allow—including the fifty-hour oil change—a figure of around $6.50 an hour. And we should assume that our aircraft will fly x hours each month—this must be realistic, as we need to know what our earnings are going to be. In a reasonably busy operation we could choose any number between 50 and 100 hours per month, so let us assume 75 hours as our estimate. (This will vary according to each person's own operation and must be checked and estimated rather closely before you begin.)

We must now allow $75 per month toward our maintenance costs (this figure may be too low—flexibility is required here) plus a further $487.50 for gas and oil. Some people like to include a sum for tires, but we'll leave that out. Our monthly direct costs are now $562.50, which when added to our fixed costs, total out at around $912.50 a month. Notice how much gasoline and oil are.

Our income will tend to vary depending on where we are in the country—higher where costs are high and lower where costs are low. But we should be able to show a minimum return of 2.5 percent on our investment. But we're not so much interested in income as in tax advantages—which is why we suggested earlier that if your annual tax is less than $2,000, you should forget about this type of operation.

Our Investment Credit allowance will be about $600 or so, and we may also claim accelerated depreciation, based on a six- or seven-year life span for the airplane. This will amount to around $3,250. We can also deduct both the interest on the financing—$500, rounded out—plus the sales tax we may have had to pay during the first year. This is quite a tidy sum, before we've even begun to consider other items, such as allowable maintenance, insurance, and so forth. You will see now why you should be paying quite a large amount of tax before you want to get into such a scheme. What you save will be able to take care of a lot of your flying expenses, while at the same time you have a nice little investment working for you.

Your second-year tax deduction will not be quite so glamorous a figure as the first, as the depreciation allowance will be down and you can only claim the sales tax once. But it is still useful. Provided you can set up the arrangement with an FBO you can trust, it can be a profitable relationship for both of you.

PARTNERSHIP

A partnership arrangement works out in much the same way as when one person is buying an aircraft, except that you divide the actual purchase price by the number of persons in the partnership. Similarly, you will have to divide the time that is yours by the number of persons in the partnership. Some airplanes are owned by partners, some of whom are *not* pilots. The only stipulation here is that one of the pilot partners does the flying for the nonflying partners. Or the nonflying partner provides his own pilot.

The most important thing about partnerships is to make sure that everything is spelled out in black and white before the purchase is made. Ideally, a proper partnership agreement is a sensible measure to which the parties will subscribe. As these agreements vary from state to state, an attorney should be sought to help draw up and file a certificate of partnership with the court clerk.

The items that need to be agreed upon include how many hours each partner will fly the airplane each month—if the aircraft is not going to be rented by an FBO—and how much the partners will charge the renter for the hours used. This may seem so obvious as to be hardly worth mentioning, but for every ten partnerships that are loosely formed by verbal agreement, a minimum of three break down because one or more members decides after the purchase of the airplane that this wasn't the way it was originally decided. So make sure that everyone agrees to everything in writing first.

Another point is to time-limit the partnership. You don't have to file anything (or write anything down, either) if the partnership is going to last less than a year. But in the case of an airplane purchase, it might be wise to time-limit the partnership to run for three years, for example, and then the airplane may be sold (or perhaps some members would want to renew the arrangement). Or you could limit it to a couple of years if you were buying a reasonably new but used airplane. Again, this is something to be worked out by the individuals.

SUMMARY

Partnerships are useful as a way to new aircraft ownership if you don't have the cash for a new airplane and feel that something older might be burdensome.

The leaseback is an excellent arrangement for those who pay high taxes and want their own personal flying to cost less.

Aircraft rental is a nice way of flying, as you don't have to settle for just one model of aircraft each time you fly. But it is usually more expensive, unless the number of hours you fly each year is small, and you don't have the satisfaction—important to some—of being able to say that you own your own airplane.

As a general rule, you can expect to get up to 80 percent of the purchase price of a new airplane financed (less for a used one) from a bank or lending institution. Banks, like insurance companies, tend to overprotect their investments in order to please their stockholders. As a result, you may do rather better from a smaller home-town bank than from one of the better-known establishments. A few banks specialize in aviation loans (as do one or two insurance companies) and are prepared to believe that pilots are generally pretty honest people. As a result, they can sometimes charge you less than other banks for the interest on your loan and can also give helpful advice on how to proceed with aircraft ownership, if that is your aim. Aircraft manufacturers are also getting into the lending business to finance sales. Interest rates, which have been keeping up with inflation the past year or so, will have to come down if the world economy is not to collapse. The Arabs, with their *embarrasse de richesse*, undoubtedly sense this problem and will hopefully reinvest their loot before it is too late for everyone.

MAINTENANCE

Aircraft are very much like cars in that they need care and attention to keep them in top running order. The sort of maintenance a sensible car owner applies to his car is similar to that of the conscientious pilot.

Checking the tire pressures, even if it's only a visual check, is a useful exercise. Those little tires rotate pretty fast when you take off, and when you land they go from zero speed to around 60 or 70 mph in a fraction of a second. Clumsy landings and high-speed turns that cause the tires to flex are what wears them out. If you can grease the airplane onto the runway, you can considerably extend the life of your tires. And if you keep them at the recommended pressures, you'll also enhance their lifespan.

Checking the battery-fluid level on some aircraft can be a bit of a hassle, but it is worth finding out where the battery is located and running a check every other week if you're not using the airplane that much. Similarly, it's sensible to run a check on all the electrical fuses, first so that you know what each one does, and second so that you know how to replace a fuse in the event of a malfunction. By the way, do you know where you're keeping your spare fuses?

You'll check the oil level every time you preflight. Still, if you don't mind spending a bit more money, it is sensible to change the oil every other month or every fifty hours, as mentioned earlier, whichever is sooner. And it also helps the engine if you use the correct weight of oil, summer and winter. Summer-weight oil is blended to provide good lubrication despite the additional high outside temperature. It is slightly more viscous than the oil you normally use in winter, which in turn has to be sufficiently thin to withstand the cold without clogging up your engine when you want to start it.

211

Most engines have some sort of oil filter, and regular changes of the oil filter when you're changing the oil can have a very beneficial effect on engine life. Every time an engine goes to work, every minute, almost microscopic particles of metal are shed from the engine shaft bearings and from other moving parts. In spite of their small size, they can cause serious wear to other parts—additional friction—if they are permitted to stay in the lubrication system. For this reason the oil flow is cycled through the oil filter, which removes these metal and other contaminants, before returning to the machinery.

The filter—there are several excellent brands around—is constructed to sift out all foreign bodies while at the same time permitting the oil to pass through it. Obviously, filters work less efficiently if they have strained out much debris; in order to keep your machinery in good order, a filter change every time you change your oil is one way to ensure a healthy engine.

Propellers also require maintenance. Unfortunately, by its very position a propeller tends to get nicks from stones and mud sucked up from the ground. Although propellers aren't that expensive compared with a new engine, their useful life can be extended by dressing out these nicks when they appear. Even the tiniest graze causes a slight imbalance to the prop when it's revolving. And although metal props are not quite as precise as the old fashioned handmade wooden ones, they are still precision airfoils, which is why they work as well as they do. Aircraft with constant-speed (or variable-pitch) propellers need to be very carefully checked.

The early designers of propellers and airplanes quickly discovered that if they were to obtain optimum efficiency from their not-too-efficient engines, something would have to be done to control the pitch of the propeller. Early methods of controlling the propeller pitch were crude by our standards, but they worked. There were two positions of the variable pitch prop, a takeoff mode and a cruise mode. Fine pitch—that is to say, relatively small forward movement—was used for takeoff, when inertia is greatest and the airplane must be made to accelerate forward. This was changed to coarse pitch, providing relatively large forward movement, when the airplane had already built up momentum for cruise. It was not long before designers noted that if they could have a totally variable rate of propeller pitch throughout every flight regime, aircraft would be more efficient. A number of automatic propellers were made, resulting today in the type we refer to as constant speed, which controls the pitch of the propeller very accurately in all flight modes.

212

The principle behind the constant-speed propeller is a very simple governor built into the propeller, that is controlled by a spring. Small bob weights within the governor rotate, and the centrifugal force they develop compresses this spring. As rpm decreases, the spring overcomes the centrifugal force of the bob weights, and the arms of the governor move inwards changing the pitch of the propeller. It is the tension of this governor spring that the pilot controls from the cockpit when he sets in a certain engine rpm that the unit will "constant speed".

The device having been adjusted, the propeller and its governing mechanism now do the work. On takeoff more power is available because the prop blades are in fine pitch; as the airplane goes faster in its takeoff run, the pitch is automatically coarsened to provide optimum efficiency. At cruise altitude the pilot reduces both engine rpm and his manifold pressure, for reasonably brisk and economic cruise flight. This coarsens the pitch of the prop blades still further, providing more thrust while using less engine power.

But suppose the pilot went into a brief dive? The pitch of the blades would decrease, and the result would, theoretically, mean an increase in engine rpm. But meanwhile the governor is speeding up, and the bob weights begin to fly outward, causing the mechanism to correct itself. The pitch is increased (coarsened) to the degree required for the prop to absorb the energy produced by the engine, thus maintaining the constant speed.

Oil is used to control the pitch mechanism by simple hydraulic action. Part of the pretakeoff procedure in aircraft equipped with constant-speed propellers, is to cycle the propeller to make sure that the oil is flowing and that the mechanism is working. Regular inspection of the unit ensures that it is in top condition; should you spot any oil leaks in this area, have the propeller inspected immediately. Constant-speed propellers have two faults. They will sometimes overspeed as a result of malfunction or loss of oil pressure and occasionally may lock into one pitch position. Cycling the propeller in flight sometimes cures the latter problem, but regular propeller maintenance is the best answer.

Fuel systems do not suffer the more obvious symptoms of wear and tear, but routine checking of the quality of fuel—plus inspection of the filter systems on a regular basis—is valuable. One of the principal sources of potential trouble is the addition of water to the fuel supply, usually through condensation. By checking the filters you can also gauge what state the fuel tanks are in. Contaminated fuel, fortunately, is almost a thing of the past.

If you do find even minute traces of water in your fuel, be especially cautious. There is almost certain to be more water in the system, and it is worthwhile checking the wing tank sumps. The container you use for checking should be clean and transparent. And if you are unlucky enough to find more water in the fuel from the sumps, it might be an excellent idea to call in your friendly A&P to have him conduct a thorough check. Although the main drain is arranged to collect water before it gets to the engine, fuel supply systems occasionally get so waterlogged (for a number of reasons—carelessness is the most common) that the whole thing has to be cleaned after draining. Fuel supply systems are quite easy to follow, and the owner's manual for your airplane will show you the items you are expected to check out as a matter of routine.

One last point about fuel. Occasionally the men or boys at the pump are rather careless about fuel. They put jet fuel into prop aircraft and vice versa. You should never rely on the pump man to put the correct grade of fuel into your airplane. Always check yourself, since it is your responsibility if you get the wrong gas. And of course you should know what is the proper fuel grade for your own airplane, which is in the owner's manual.

Aircraft fuels are colored to indicate their grade. The color coding is as follows:

Red fuel: Aviation grade 80/87 octane
Blue fuel: Aviation grade 91/96 octane
Green fuel: Aviation grade 100/130 octane
Purple fuel: Aviation grade 115/145 octane
Colorless or
   straw colored: Jet A grade fuel

As mentioned earlier, if you can't find the right fuel and you need some to get you to an airport elsewhere, use a small quantity of the next higher grade. On no account use a lower grade. It is sometimes helpful to have a chamois leather to use as a fuel strainer if you're uncertain about the purity of your fuel source. The chamois will filter out the water that may be present. One final point, gasoline weighs six pounds per gallon.

Of the instrumentation checks, perhaps the most important is the VOR receiver check, as sensitivity and accuracy are vital to safe navigation on instruments. (I'm assuming that you will regularly check the altimeter—not more than 75 feet in error). The FARs require a check of the VOR's accuracy every ten hours of flight and within ten days of an IFR flight operation. You are required to keep a permanent record of this accuracy check in the aircraft log book, giving the date when the check was carried out, the place where it was carried out, any bearing error, and the signature of the person who actually conducted the test. Part 4 of the AIM gives details of suitable test locations, both on the ground and those usable from the air. The following acceptable checks are listed in order of preference:

1. VOT—VOR test system. Not an omni station, the VOT is purely for testing and does not make use of a landmark. You test by setting the OBS to 360 degrees with a FROM indication and by setting the OBS to 180 degrees with a TO indication. Allowable tolerance is ±4 degrees.

2. Designated ground checkpoint. You'll find full details in the AIM of how to conduct this type of check, and you should follow the instructions carefully. Park your airplane in the test area and set the OBS to the published radial. Allowable tolerance is once more ±4 degrees.

3. Designated airborne checkpoint, as found in the AIM. Once more, follow the details carefully, setting the OBS to the published radial, and fly over the designated landmark. Tolerance for a published airborne check is ±6 degrees.

4. Your own airborne check may be set up as follows. First, select a landmark that lies on a published VOR victor airway, preferably at least twenty miles from the VOR you'll be flying toward. Set the OBS to the predetermined radial, and fly over the landmark at a reasonably low altitude. Tolerance for this type of check is also ±6 degrees.

5. Perhaps the easiest of all the checks is the dual VOR check. Set both receivers to the same station with centered needles and a TO indication. Now check the OBS settings. Tolerance here is ±4 degrees for each set.

One last point you should remember when conducting VOR receiver checks is that a localizer course is not acceptable as a VOR check. As with the mechanical side of aircraft maintenance, so with the avionics—maintain a regular system that will ensure that everything works when you need it.

There's one item that we haven't yet covered. In one way it really should have been the first item mentioned when we began with the problem of maintenance: yourself. Unless you are keeping yourself reasonably fit, you are not going to be able to enjoy the pleasures of flight to their fullest extent. And if you aren't in reasonably good shape, you could be operating at less than maximum efficiency. We should try to be kind to our bodies, to look after their wants, and to have them checked regularly. We cannot function at our best if we are overweight, underexercised, or overtired. And we cannot function as pilots if we don't keep ourselves in reasonable shape.

So keep a check of your own health. If you've got a cold, get rid of it before you get into that cockpit. If you're feeling under the weather from a little too much celebration, wait a day or two until you're less fatigued. For if you are not in top condition yourself, how will you be able to look after your airplane to the best of your abilities? And how will you best be able to enjoy all that flying can afford you?

The FARs dealing with aircraft maintenance, preventive maintenance, and alterations are covered in part 43, "Maintenance, Preventive Maintenance, Rebuilding and Alterations," and in subpart C of FAR part 91, section 91.161 and the sections that follow.

10

Radios and Avionics for Lightplanes

If you don't want to have a radio in your airplane, there's nothing that says you must have one. Of course, you won't normally be able to land at fields that require a two-way radio in airplanes. But as many a ferry pilot can tell you, countless flights are made each year in aircraft without radios. And if you should have to go to a field where a two-way radio is required, you can usually get permission by calling the tower by phone, giving them an ETA and asking if they'll give you a green light to land. Depending on the traffic, you'll usually find them helpful and accommodating.

At the present time there is considerable argument between aircraft owners and the FAA about the FAA planners' decision to change band communication from today's 50-kHz spacing between bands to 25-kHz. A lot of bureaucratic empire-building is involved in the decision, since the FAA itself has been rather free with its use of the 50-kHz channels and has started to use them for Automatic Terminal Information Service (ATIS) broadcasts. Until quite recently this information was broadcast over the voice feature of radio navigational aids, but the bureaucrats decided that "some pilots" objected to the background noise and consequently have allocated more than 100 locations to use the 50-kHz discrete communication frequencies. As the cost of the changeover to the 25-kHz spacing is going to cost general aviation some $400 million, the argument could become quite heated.

Furthermore, the overall increase in air traffic, though considerable, is in no way as serious as the government would have us believe. If every single airplane in the United States were to be up in the air at the same time, there would still be a separation of some twenty-five miles between each one. It is only when you get a lot of aircraft centering upon a particular area that congestion occurs. And the principal cause of congestion has been the airlines, who out of the more than 12,000 airfields within the forty-eight contiguous states, limit their flights to some 750, only serving some 22 cities frequently. The oil "crisis" has served to limit further the airlines schedules.

RADIO: WAVES AND WAVELENGTHS

The best way to understand radio waves is to think of them as being a continuation of the spectrum of waves known as light. Depending on our eyesight, we can view a source of light on a clear day (or night) from a considerable distance, until such time as the curvature of the earth interrupts the path of the light from its source to our eyes. The same thing happens with very high frequency (VHF) waves. In a further experiment, we can beam light so that it shines only in a certain direction. If we now attach a shutter to the beaming device, we have the sort of signaling equipment that is found in ships for nonradio communications, and the VOR works on a very similar principle.

Light is a matter of wavelengths, at least as far as we can perceive it—at the present time we have not evolved a sufficiently sophisticated organ to perceive the greater part of the electromagnetic spectrum of which light forms a small portion. It is bounded at its lower end by infrared radiations—invisible to our eyes, they can be used through electronic equipment to see through fog or darkness—and at the upper end by ultraviolet waves, which are very damaging to our eyes but beneficial (in small doses) for our skin. Between these two extremes falls what is termed visible light, which may be broken into specific radiations we term the colors red (longest wave length), orange, yellow, green, blue, indigo, and violet. The actual frequencies involved here range from 1,000 Gigacycles per second for the bottom end of the infrared portion to 100 million Gigacycles at the top end of the ultraviolet part; the actual wavelengths involved are from approximately 0.5 mm down to 0.000005 mm, respectively.

Gigacycles are simply a convenient way of measuring frequencies. Immediately below the infrared frequencies lies a gamut of frequencies that we use for radio communication. At the bottom end of the spectrum lies electricity itself. Toward the top end of the scale are X-rays. All this is part of an extremely large portion of universal energy, of which radio waves form a sector. And like light itself, radio waves may be beamed in specific directions or they may be non- or omnidirectional.

So far no one has succeeded in photographing radio waves, but mathematicians have given us a good clue as to how they appear to work. Electromagnetic (radio) waves are totally balanced forms of energy, fluctuating from a point zero to a maximum intensity, then declining through the point zero to repeat the sequence to a mirror image maximum in an opposite polarity completing a cycle. The total process is measured by time from the beginning of the cycle through the completion of its equal and opposite parts. (See Figure 10.1.) This time-distance is called a wavelength; the length of each

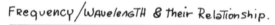

Frequency/Wavelength & their Relationship.

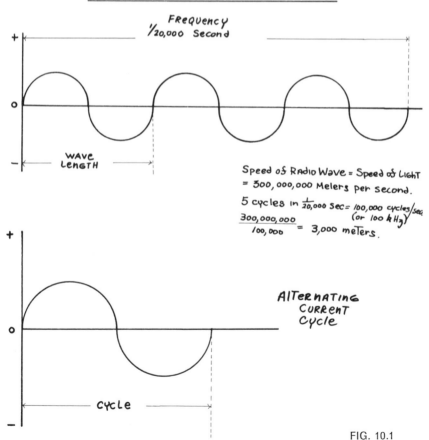

Speed of Radio Wave = Speed of Light
= 300,000,000 Meters per second.
5 cycles in $\frac{1}{20,000}$ Sec = 100,000 cycles/sec.
(or 100 kHz)
$\frac{300,000,000}{100,000}$ = 3,000 meters.

FIG. 10.1

219

cycle measured in meters is a radio wavelength. This same measurement expressed as the number of cycles that occurs in each second is frequency. As far as aviation is concerned, frequency describes that portion of the waveband that a radio facility is using.

Household current in the United States is usually supplied at sixty cycles per second; where it is not properly grounded it can be heard as a hum on television and radio. Normal radios, however, receive signals at much higher frequencies than the domestic sixty cycles, and we therefore refer to these frequencies in thousands of cycles, or kilocycles, per second and also as kiloHertz per second. For very high frequencies we use megacycles or megaHertz (millions of cycles per second). Finally, we even use Gigacycles when we are speaking of thousands of megacycles per second. Both the FCC and the FAA have actually adopted Hertz as the basic unit of frequency as opposed to cycles: The reason is to honor the German scientist, Heinrich Hertz, who is generally credited with the discovery of radio waves.

Converting from frequency to wavelength or from wavelength to frequency is very simple. Divide either measurement into 300,-000,000 to obtain the other, 300,000,000 meters being the distance light (and radio waves) will travel in one second (approximately seven times around the earth).

For convenience, and since we use different parts of the radio wave spectrum for different purposes, radio waves are subdivided into frequency wavebands. Each band has its own peculiar properties. The Low, Medium, and High frequencies cover from 30 to 300 kHz, 300 to 3,000 kHz, and 3,000 to 30,000 kHz, respectively. All three bands are susceptible to interference from natural radiation, such as sunspots or thunderstorms. Very High Frequencies, run from 30,000 to 300,000 kHz, or more simply from 30 to 300 MHz and are remarkably free from static interference. But they have a different problem—shortness of range, especially when used between ground stations, where they are limited to line-of-sight range. The High Frequency waveband, on the other hand, is fine for long-range communication, say from Washington to Peking or from London to Sydney, Australia—but only when conditions are good and there's not too much radiation.

One of the reasons for the differences in the behavior of radio waves is the existence of two layers of electrically charged gases surrounding our atmosphere. The upper layer is known as the Appleton layer, the inner as the Kennelly-Heaviside layer. Both reflect back radio signals of a certain wavelength and intensity that vary according to frequency. By bouncing back waves of a certain

220

length, long-distance radio communication becomes possible. Very Low Frequencies creep across the world's surface and may even be detected under water by submarines. Meanwhile, the waves of Very High Frequencies pass through the two layers and go on into outer space—they don't get reflected back, which is what limits them to line-of-sight range.

Normal television stations make use of very high frequencies for their transmissions. Ultra high frequency (UHF) wavelengths are used mainly for very short-range communications but may also be used in conjunction with satellites that can rebeam the signals to where they are wanted. Radar uses super high frequencies of 3 to 30 Gc/s; at the present time the extremely high frequency (EHF) band from 30 to 300 Gc/s is listed as experimental.

The radio transmitter is a device that produces a signal amplified for transmission. The receiving part of our radio works by reversing the process; it accepts the signal and breaks it down into sound. The two keys to radio communication are accurate antenna tuning —the more closely the antenna length matches the wavelength being used, the better the reception—and more accurate frequency control. Accurate frequency control at both transmitting and receiving points ensures distortion-free communication.

VHF communication depends on frequency modulation for its effect, in exactly the same way that FM radio stations produce undistorted sound. What happens is that sound in all its complexities of tone and pitch is collected by the microphone, which changes the air vibrations caused by our speech into electrical currents. These currents are then "printed" on the radio carrier wave being generated by our transmitter, which modulates its frequency in relation to the input from the microphone. This process is reversed by the receiver, and a pretty reasonable facsimile of our voice and what we're saying comes out at the other end.

COSTS

Having dealt with the theory part of it, now let's see what we're likely to need in the cockpit. There's no doubt that a radio in an airplane greatly increases its utility, and in certain respects can be said to enhance safety. Most pilots, unless they're planning to fly instruments most of the time from the beginning, work up to a full radio panel in stages. The principal reason is cost; it is easy for the radios in an aircraft to constitute one-third of its overall new cost, and it is not at all difficult for the radios and navigational equipment in a secondhand single-engined aircraft to be worth nearly as much as the aircraft itself.

If you are equipping a used airplane, you should first check which of the normal instruments (not radios) are powered by electricity and which work off the vacuum. You should do this before laying out any money, as it is not unheard of for aircraft salesmen to doll out a used airplane with a flashy full panel that actually isn't worth too much if you need it for instrument flight. Your A&P can advise you if you don't know where to get started, but ideally you want these instruments split down the middle, half operating off vacuum, the others on electricity, so that if you get an electrical failure you have sufficient instrumentation to get home. At the present time, artificial horizons can be both electrically or vacuum driven—check to see which in your airplane—while the DG is vacuum driven. The turn and bank, or turn coordinator, is usually electric; the little compass in the windshield is your backup for the DG. Many pilots install the Kollsman Direction Indicator, which is simply a larger wet compass, carefully balanced to vary less around the poles. There are also electric compasses that can provide magnetic compass headings based on the airplane's distant reading compass and that may also be used to get bearings from radio beacons (both ADF and VORs). These can be very expensive.

Having made sure your normal instrumentation is trustworthy, you now start to think about a radio. You'll probably decide that a simple navcomm is the very minimum you can do with. And with the almost certainty of increasing regulation by the FAA of those frequencies, you'd better think in terms of at least a 200-channel transceiver. If you know of a good radio shop, it is sometimes possible to buy good secondhand equipment that has been checked out; in this way you might be able to pick up a 360-channel set immediately.

If money isn't too much of a problem, you should probably go straight to a 720-channel set, which starts at just over $1,000. Another possibility would be to get what used to be called a 1½ system, which is a unit in which both nav and comm share an outer covering and in which either the nav or comm may be used for transmitting. Today's sets allow a VOR to be tuned in and to work while the comm part is available for messages. If you are going to use the 1½ system, it is sensible to get two of them, although it can sometimes be a puzzle as to which you are using at one time if you don't have some sort of panel so that you can switch sets in and out. Actually, anytime you have more than a single set it's worth having a switching panel!

FIG. 10.2

Narco's Mark 16 replaced the popular Mark 12 series as an efficient, moderately expensive navigation and communication's transceiver. This set is the minimum for IFR usage.

FIG. 10.3

The 1½-system is the poor person's method to inexpensive radios. The set includes a communications and navigation unit, but you only get to use one at a time. While set to COM, the navigation unit retains some sensitivity.

FIG. 10.4

Omni head with glide scope bar

As far as the nav receiving system is concerned, you may or may not have to buy a separate omni head. If you do, it's worthwhile getting one that matches the receiver. Also, make sure that the nav receiver has localizer channels, and that these will be displayed by the omni head. If you are planning on flying instruments, you can manage with this minimum of equipment if you have to.

If you do intend to do much instrument work but are limited by cash, another possibility is to get an omni head that features a glide slope bar and a nav receiver that offers a glide slope too. Although a number of people fly instruments with just one transceiver, it can be tricky. Two sets make instrument flying much easier and give you standby capability too. Only one set for IFR is limiting, since if your set should break down while you're flying in bad weather, you'd better have a convenient out. Don't forget to check that single omni for accuracy every chance you have.

If you'll be flying in out-of-the-way places, you may think in terms of an ADF instead of a VOR. But the country is so dotted with VORs that unless your flying is in really desolate areas, your ADF will probably be relegated to your budget. ADFs can be used with localizers and the digitally tuned units now available, are easier to use than the old-fashioned knob-twiddling variety. Think about it—in terms of what gets switched off first.

If you live in any of the busy aviation states—those covered by the eastern, great Lakes, southern, and western regions of the FAA (and Texas, of course) and you intend to pass some of your time in and around terminal areas, a transponder is going to make more and more sense. The FAA radar sets are not too good at picking up aircraft without transponders.

The transponder is an outgrowth of a device developed during World War II to distinguish friendly from hostile aircraft. A ground station sends out a signal that activates a black box carried in the aircraft, triggering a response that identifies the aircraft as a friendly one. For security reasons—the device was developed and in use by the time of the Battle of Britain—it was known only as Identification: Friend or Foe (IFF), but today it has found peacetime use in permitting the discrete identification of aircraft on radar screens. Transponders are now hooked up automatically with aircraft altimeters, which gives the air traffic controllers the read out of an individual aircraft's altitude and flight numbers of airliners. (Naturally, this will make it more important to fly at exactly the right altitude!)

Transponders cost from $495, and even some of the cheap ones have provisions for the altitude-reporting altimeter. It seems very

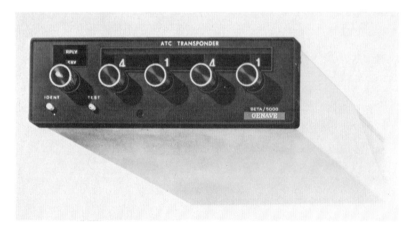

FIG. 10.5

Digital tuning transponder helps make life easier for ground-based air traffic controllers. Ground control, when seeking to identify your aircraft, will request you "squawk ident" on a code given. This means pressing the ident button (lower left), which signals your airplane as two slashes on the controller's radarscope.

probable that altitude-reporting altimeters will become mandatory on all aircraft filing IFR—once the FAA buys the necessary equipment for itself—and will almost certainly be a requirement for aircraft operating within TCAs (Traffic Control Areas). It makes good sense, therefore, to buy a transponder that can handle this sort of task.

We'll assume that you now have a navcomm with omni head and a transponder—and possibly an ADF. This is really all you need for

FIG. 10.6

Bearing indicator for Sigma/1500 ADF

FIG. 10.7

Solid state circuitry heightens reliability of modern avionics, reduces
weight and cost. Shown here is the
Genave Delta/303 marker beacon
receiver.

most flights, but if you can afford it and if you are intending to fly
quite a lot of IFR, perhaps the next item you might want to add
would be a marker-beacon receiver.

The localizer and the glide slope keep the pilot informed that he
is in line with the runway and descending at the proper rate.
Marker beacons basically give the pilot a progress report of the
distance to touchdown. At most civil airports there are two markers: an outer one, which is the most distant from the airfield, and a
middle marker, which, site permitting, is usually less than a mile
from the runway threshold. The actual distances are quoted on the
respective approach plates for each airfield. Inner marker beacons
are used by the military and are not necessarily found at civil
airfields.

Marker-beacon receivers are tuned to a single frequency of 75
MHz and will respond to signals at 400 cycles (outer marker) and
1,300 cycles (middle marker). It will also respond to signals of
3,000 cycles (inner marker), which frequency is also used for fan

markers, found on certain airways where accurate position reports are required by ATC.

Flying over a marker beacon, the appropriate circuit within the receiver is activated by a ground signal. At the outer marker, for example, the blue light with OM written by it will flash a series of dashes and a low-pitched audio signal will be heard. Continuing on approach, the middle marker identifies itself with an amber light in the panel, which flashes a series of dots and dashes, together with a medium-pitched audio signal. Fan markers activate the third light —a clear bulb—which flashes a series of dots on station passage, with a high-pitched audio signal.

Another relatively inexpensive item you might want to add with a view to the future is the audio-switching panel just mentioned. Some of these have marker-beacon receivers built into them, which is very useful, as this can save you money by obviating a separate installation. This panel enables you to select either transmitter and any number of receivers you may want to monitor at one time. It's a convenience item to be sure and isn't absolutely necessary, but you'll come to love that convenience once you start using it.

Emergency locator transmitters run from $90 to more than $500. A typical one, weighing 2.3 pounds, would cost about $185, which provides you with permanent installation, impact operation (if you hit the ground in a crash at more than a certain force, it sets itself off automatically sending a distress signal), a test switch so that you can check if it works, and the possibility of using it as a portable set, with manual operation and with voice modulation on the distress frequency, of 125 MHz.

The least expensive communications transceiver with 360-channel spacing would cost $600, which gives you 50-kHz capability. The least expensive navcomm unit, with 50-kHz spacing on both nav and comm, would cost about $1,000, which would include an omni head. For that you'll get 560 channels (a 1½ system). The least expensive 720-channel set is just short of $1,200 and includes a 200-channel nav system within it.

The 560-channel set includes a glide slope and the electronics required for Distance Measuring Equipment (DME) switching. A marker-beacon receiver would cost an additional $130.

If you wanted an ADF, the least expensive digitally tuning model is just under $1,000 and includes a digital stopwatch device, which is very handy for instrument approaches. An inexpensive transponder that has encoding altimeter coupling costs around $600. This unit has a standby switch. Encoding altimeters are still a bit costly, so you might decide to wait until the prices come down.

The least expensive costs $600 and prices go up much higher. There is a very simplified DF system that uses the VHF communication receiver, allowing you to steer according to signals from Unicoms, towers, and so on. These cost from around $500. I've never used one and don't know how good they are, but there's no reason why they should not be useful.

LIGHTING AND HEATING

At this point you might feel that you'd like to think of anticollision lighting for your airplane. All aircraft with an original-type certificate dated later than August 11, 1972, must have a light of 400 effective candlepower. FAR 91.33 spells out what the requirements are for everyone else. What all this means is that you must have a light that will shine through 360 degrees around the vertical axis of the airplane and both 30 degrees above and below the horizontal plane. This light may be of 100 candlepower. That's not a lot, even on a clear day, and many pilots like to introduce strobe equipment to make sure they are seen clearly.

The cheapest strobe replacement of the old incandescent flashing lamp (on top of the tail) costs $85. Candlepower is 125, and the device flashes sixty times a minute. The same company produces a similar model to replace the rotating beacon for $100, and this produces slightly more candlepower—150 effective candlepower to be precise. A three-light kit, which will cover wing tips and the tail, will cost, depending on your aircraft, between $300 and $400.

In the interests of fuel economy, you might at this point consider the installation of an exhaust gas temperature gauge (EGT), which is not expensive. Pilots who use them assure you that they pay for themselves in less than a year in gas saved. Starting price is about $100 and about $200 if you want to monitor all the cylinders— worth the price, by the way, not just in saving gas since an EGT can detect most incipient combustion chamber malfunctions well before they occur. When you can spot something going wrong in a cylinder at an early stage, you can check what's causing the trouble and avoid one of those landings without power.

MICROPHONES

In the interests of your own aircraft handling, you might think about replacing those ridiculous hand-held microphones with something else, such as a boom mike. This really makes a lot of

228

sense, because if you're on instruments and rather busy, you don't really need to be searching for the mike to make a position report. Boom mikes also make sense even if you don't fly instruments, since a boom mike permits you to keep flying the airplane while you're talking. The cheapest versions cost about $40, including a noise-cancelling feature. Another possibility would be to get a headset. An inexpensive headset with volume control will run you from $20 to $25. You can also get one of those super-lightweight combined headset/mike devices. The general aviation version of the type the NASA people use costs about $100.

AUTOPILOT

At about this point, your thoughts may run to having some sort of autopilot. If most of your flying is VFR, an autopilot may be unnecessary, unless you really feel your time is better occupied with keeping a very good look around outside for other traffic. (You could also consider a proximity warning indicator.) Simple autopilots keep the wings level; they may also include options for holding heading too. Expect to pay under $900 for one of these, which will give you roll and yaw control and heading hold. For another $125 you can get VOR and localizer tracking. A full three-axis autopilot costs rather more, and one that includes an altitude-holding device with crosswind compensation costs about $4,000, which includes VOR and localizer tracking, of course.

Now for your second navcomm, assuming you didn't already get it. With dual navcomm and transponder, plus your glide slope and marker-beacon receiver, you can handle most IFR work. If you will be doing a lot of instrument flying, you may want to consider investing in Distance Measuring Equipment (DME), a device that tells the pilot the distance he is from a VORTAC and also at what speed he is approaching it.

DME can be used in conjunction with other aids but is usually associated with VORTACS. It provides you with bearing and distance, presented in digital form on most sets. Because an airplane is above the surface of the earth, the actual readout is really slant-range, which varies depending on the height of the aircraft. It is always greater than the actual ground distance would be. Think of slant range as the hypotenuse of a vertical triangle as in trigonometry, with the ground distance the horizontal leg.

The problem with DME is that the number of aircraft now using it is growing, and ground facilities have at present a saturation

FIG. 10.8

DME's provide distance measuring and speed of aircraft in relation to a ground beacon. Because of the altitude the distance measured is slant range—but that's accurate enough when combined with time to target.

level of around 2,960 inquiries per second. DME could be said to work a bit like a transponder in reverse—it queries the ground station and by measuring the ground station's signal can work out the distance and speed at which it is approaching that station. The range of most DMEs is about 200 miles, with an accuracy of about plus or minus 3 percent. The least expensive DME at present is $1,600—this set has a range of 100 nautical miles and uses a needle rather than a digital readout.

As far as lightplanes are concerned, there is a point at which adding more equipment is counterproductive in terms of useful cabinload. Of course, many corporate aircraft have as much and as good equipment as some of the biggest airliners. (About the only thing you won't find on corporate aircraft at the present time is an inertial navigation system. What you choose to buy now is very much up to you. You could invest in area navigation (RNAV), which is basically a small computer that allows you to move VORs to where you want them to be for your journey. ATC is beginning to understand that RNAV's only threat is to lighten their workload, so they are getting used to it.

If you already have DME, you could add RNAV for about $2,000. Prices will probably come down. Or you could do what some people do in larger aircraft—buy yourself a Flight Director system.

The Flight Director assembles all the pertinent information for your flight and presents it on a single instrument for you to read. It even tells you what you should do, if you feed in the data beforehand. Some have autopilots tied in, others have automatic pitch stabilization, and several offer a SCRAM button should you want to abort a landing. For example, just press the switch and the airplane automatically adds power, takes itself up six degrees, cleaning itself up as you're on your way. Inexpensive flight directors start at about $4,000.

Business tycoons often add telephones to their aircraft. The cheapest air-telephone costs about $2,000; one call costs a minimum of $1 plus the cost of the ground portion of the call. If you're a business executive, no doubt you could have a WATS line arrangement with the telephone company and include your airplane's radio/telephone in the arrangement.

With Congress about to act on its own, the FAA is expected to rule very shortly on what type of collision-avoidance system it will promote. At the present time several methods are available to alert the pilot, but only one system is ready for marketing. The problem is to what do you make such a system respond—transponders, strobes, infrared emissions, or what? What there is at the present time costs about $400; if you feel your eyes aren't that good when it comes to spotting other aircraft, this equipment might be for you.

Inexpensive air conditioning costs less than $1,000. Perhaps its biggest drawback is weight; none of the sets presently offered weighs less than 60 pounds. If you don't have that much leeway in useful cabinload in your airplane, you might want to forget about it.

SURVIVAL EQUIPMENT

Survival equipment is something most pilots never bother to think about. But the United States is a large place, and even the most apparently densely populated areas are still situated near small areas of wilderness. Not too long ago an aircraft disappeared for more than six months just five minutes out of suburban Connecticut.

Survival equipment and what it should contain is a matter of individual choice. If you do much flying over water, even on short hops, it makes very good sense to carry at least life jackets for you and your passengers. (If you make long over-water trips it's required.) If you make extended trips, you may want to add other equipment, such as a life raft. Don't let the weight factor put you off—this is one time when it's better to be safe.

Another very useful item to have on board is oxygen. Even if you don't intend to do much flying above 10,000 feet—and if the FAA has its way, you probably won't—having oxygen on board can make all the difference at the end of a longish night flight at 8,000 feet, for example. Taken for half an hour before landing, you'll be amazed at how your night vision improves and in how much better shape you'll be for the landing. If you can spare $125 for oxygen equipment, you'll get it without an automatic regulator, it is true, but with two outlets, which should be all you'll need. That would give you more than three and a half hours' oxygen for two persons.

Weather radar is still for two-engine aircraft, although several persons are working on single-engine weather radar systems. Price is also rather high, considering the advances that have been made in circuitry technique. Once single-engine radar becomes a reality, it seems likely that there will be an overall reduction in price. What seems to be keeping the price up at the present time is the type of scanning system still in use; electronic scanning will effectively reduce the cost. Present weather radar systems start at around $7,000. Single-engine radar, mounted within the leading edge of the wing, should not cost much more than a current good ADF.

There are many other items available for making your cockpit comfortable. Some people make a fetish of kneeboards, super Swiss stopwatches for timing, map lights for reading, even special panel lighting (sometimes a very worthwhile addition). You may want to provide your rear-seat passengers with a full harness, which they really should have. Even a lowly fire extinguisher is a useful addition; if you get one, make sure it is capable of dealing with both electrical and other types of fire and that the chemicals it uses aren't poisonous to human beings. One can also install a stereo tape machine which will not merely provide stereo entertainment for those long hauls but also can record and play back clearances from ATC.

Perhaps the most important thing to remember when you're outfitting your airplane is not to buy anything unless you've established a need for it. Then find out how much it weighs.

RADIO LICENSING

Unlike other countries where radio operators of all types of radio are required to take an examination to show that they know how to operate the equipment properly, the FCC requires only two documents. The first is a restricted radio-telephone operator's permit. This is obtained by filling out the appropriate application form and sending it with your check to the nearest FCC field office.

The second item is a radio station license; with a transceiver in it, your airplane is treated as a mobile radio station. This document must be in the airplane whenever the transmitter is used. (It is your responsibility to see that these rules are complied with.) This station license is good for five years. If you are applying for such a license for the first time, after buying either a new or used airplane, you have thirty days in which to make application in your name.

FREQUENCY ALLOCATION

At the present time the FCC is revising frequency allocation, but at the time of writing the following were in force.

Air Navigation Aids

108.1 to 111.9 MHz	Localizer with simultaneous radio-telephone channel, operating on odd-length decimal frequencies (108.1, 108.3, 108.5, etc.)
108.2 to 111.8 MHz	VORs operating on even-length decimal frequencies (108.2, 108.4, 108.6, etc.)
112.0 to 117.9 MHz	VORs for airways track guidance

Communication Frequencies

118.0 to 121.4 MHz	Air Traffic Control
121.5 MHz	*International emergency frequency*
121.6 to 121.9 MHz	ATC—ground control
121.95 MHz	Flight test
122.0 MHz	FSS—en route weather service (flight watch)
122.05 to 122.15 MHz	FSS simplex or receive only with VOR
122.2 MHz	Available at all FSS for en route service
122.25 to 122.45 MHz	FSS simplex
122.5 MHz	Tower receive only
122.55 to 122.75 MHz	FSS simplex
122.8 MHz	Unicom—airports with no tower or FSS
122.85 and 122.95 MHz	Unicom—available for private-use airports
122.9 MHz	Multicom (agricultural, forestry, ranching, fire-fighting, parachuting) and air-to-air
123.0 MHz	Unicom—airports with tower or FSS
123.1 MHz	Search-and-rescue and temporary tower
123.15, 123.20, 123.25, 123.35, 123.40, 123.45, and 123.55 MHz	Flight test
123.3 and 123.5 MHz	Flying schools
123.6 and 123.65 MHz	FSS, airport advisory service
123.7 to 123.95 MHz	Air Traffic Control
124.0 MHz	Corporate communications (discrete)
128.85 to 132.0 MHz	Air Traffic Control communications (IFR traffic primarily)
132.05 to 135.95 MHz	Air Traffic Control

At the present time the ground control frequencies most commonly used are 121.9 and 121.7 MHz as the secondary frequency. Also used where interference with other stations may occur are 121.8 and 121.6 MHz. Ground control is intended to provide information to aircraft on the ground, either prior to takeoff or following landing, plus taxi instructions to aircraft on the ground. On arrival at an unfamiliar airfield, ground control will tell you where to go to tie-down, to fuel, and so on.

Control Tower Frequencies

Towers may have more than one frequency. Both the AIM and up-to-date charts carry details. At larger airports the sort of information that towers provide is handled via ATIS over the voice feature of VORs, ILSs, and so forth. Before arrival you should tune in for information, which will include the following:

1. The name of airport plus a broadcast identifier (for example, say that you have received information bravo).
2. Current weather information, which may include details of altimeter setting and temperature.
3. Instrument runway in use.
4. Landing and takeoff runways.
5. Notams, airman advisories, and any unique information—for example, if a runway had just been ploughed of snow you might be informed that braking action was still not very good.
6. Frequencies for approach and departure control.

Both approach and departure control use radar and will give radar vectors and radar traffic advisories. Traffic information from radar controllers is described in clock position, in which case the nose of your aircraft is twelve o'clock. Thus, an aircraft to the left of your airplane and at right angles to it would be described as at your nine o'clock position. If it were on the right, it would be at your three o'clock position.

Bear in mind that radar controllers are not able to see any aircraft. They are viewing electronic signals on a screen on which some light aircraft (especially wooden ones and gliders) may not show up too well.

Flight Service Stations

Of special interest to the VFR pilot is the FSS system, which is helpful in giving all sorts of information from en route weather to local knowledge helpful on cross-country flights. It may also be used for opening and closing flight plans.

Both airplane and ground stations use frequencies of 122.6 and 123.6 MHz, the latter frequency usually being for airport advisories at fields at which there is no control tower. The most usual frequency is 122.1 MHz for aircraft to transmit on, while receiving on the voice feature of a VOR. The navigation chart and the AIM gives details.

When calling up an FSS be sure to say on which frequency you are transmitting and on which frequency you are listening, as FSS frequently monitor several stations on different frequencies. And don't forget EWAS, the discrete weather frequency at 122.0 MHz; it is *only* for in-flight weather briefing.

Unicom

The Frequency of 122.8 MHz is most usually assigned for communications (air-to-ground) at airports at which no control tower or FSS exists. Unfortunately, it is an overburdened frequency. At airports with control towers the unicom frequency is 123.0 MHz. Unicom is used for fueling, food, transportation, and even personal messages.

Air-to-Air

The frequency of 122.9 MHz is a multicom frequency used by agricultural pilots, ranchers, fire fighters, parachute jumpers, and so on. It may also be used for air-to-air communication.

USING YOUR RADIO

Making the best use of radio seems for some pilots to be more difficult than it really should be. Part of the problem lies in the fact that many pilots don't understand that if more than one person uses a waveband and someone else tries to come in while the first person is still talking, that second person doesn't get heard. Rather, an electronic squeal is heard, which tends to garble the first person's transmission. So before pressing the button to speak, make sure there's no one on the waveband you intend using.

Another point to remember is to press the button down just prior to speaking. Many pilots press down the button as they begin to speak; the result is that the first part of the transmission is clipped. Another good practice is to keep your mouth close to the mike— this keeps out extraneous noise—but you don't need to shout.

Sending Numbers

Numbers are occasionally misunderstood on the radio, and for this reason a technique has been adopted to make them easier to follow. (You will notice that in sending the call sign for your aircraft, you should pronounce the number 9 as "niner." This is to avoid confusion with the German word "nein," which means "no.") Multiple-digit numbers are broken down into their single-digit form. Aircraft N3674U would give its call sign as "November three six seven four uniform." Radio frequencies are also broken down; the frequency 122.65 MHz, for example, would be given as "one two two point six five" or as "one two two decimal six five." A low frequency, for example, for a radio beacon, would add the words kiloHertz, as in "two four seven kiloHertz" for 247 kHz.

There are some exceptions. Victor Airways are usually read as "Victor Fourteen" for Victor 14. Altimeter settings usually drop the decimal: "altimeter setting three zero zero five" for 30.05, for example. And round numbers, especially for ceilings and altitudes, are read as "one thousand, five hundred" for 1,500, up to "niner thousand niner hundred" for 9,900. For 10,000 and upward read the digits preceding the thousands place separately: 16,500 would be "one six thousand five hundred."

The Phonetic Alphabet

The phonetic alphabet used today was adopted from the NATO phonetic alphabet and reads as follows:

A	Alpha	N	November
B	Bravo	O	Oscar
C	Charlie	P	Papa
D	Delta	Q	Quebec
E	Echo	R	Romeo
F	Foxtrot	S	Sierra
G	Golf	T	Tango
H	Hotel	U	Uniform
I	India	V	Victor
J	Juliet	W	Whiskey
K	Kilo	X	X-ray
L	Lima	Y	Yankee
M	Mike	Z	Zulu

Zulu, or zulu time, refers to Greenwich Mean Time.

Telling the Time

We have already discussed how important a part time plays in the business of a pilot. Flight plans are always filed using the twenty-four-hour clock system, most usually according to Greenwich mean time.

Under the twenty-four-hour clock system, the first two digits are used to show the actual hour and the last two digits show the minutes after that hour. One o'clock in the afternoon would be described as 1300, six o'clock in the evening as 1800 and so on.

Air traffic control works according to Greenwich Mean Time, and air carriers around the world schedule their aircraft accordingly. It is worth getting used to, although ATC also works on local time for general aviation aircraft. If you want to know what the Greenwich time is as compared with local time, the following information will assist you: To obtain the Greenwich mean time, add 5 hours to Eastern Standard, 6 hours to Central Standard Time, 7 hours to Mountain Standard Time, and 8 hours to Pacific Standard Time.

11

Instrument Flying

Noninstrument pilots should not allow themselves to be intimidated by self-appointed experts on instrument flying but should try to assess the pros and cons of an instrument rating calmly.

There are very few cons; perhaps the only one of importance is the question of staying current when you can only manage a relatively small number of hours each year. One way to deal with this problem is to make every flight an instrument flight in order to keep abreast with the regulations. Unfortunately, if you have ATC holding your hand all the time, your ability to fly VFR can get impaired, strange though that may seem. Another way of tackling the problem, is to have a twice yearly refresher, partly in a simulator, partly with an instructor.

The pros of being instrument rated are many, not the least of which is the advantage of being able to fly even when the weather is keeping the VFR-only pilots on the ground. Another advantage is that the entire air traffic system has been geared to the twenty-four-hour, 365-days-a-year requirement of airliners and international traffic. Your instrument rating enables you to take full advantage of this air traffic system regardless of weather or time of day year-round.

Perhaps the biggest difference between instrument flight today and in the past is that there's so much more to know. The requirements in book learning call for a certain amount of application. It is perfectly possible to continue at flight school, completing the private certificate and then simply building up the hours necessary for both the commercial and instrument tickets. Opinions differ as to whether or not this is actually the best way. Some say yes: The

238

student can apply the principles he has already learned and by working on them with access to fairly immediate information from his instructors can perfect his skills. Others suggest that it is better for the new private pilot to get away from flight school and fly on his own, improving his skills by the mistakes and traps he will almost certainly create on his own. Then, having successfully overcome them, he should return to his tutors, correct any errors in technique he may have possibly acquired during his time away, and work on the commercial and instrument examinations.

Proponents of the first method point out that by leaving the tutoring atmosphere the pilot may become lazy. This may be a hazard, but on the other hand, anyone who really intends to learn something has only to want that learning to be able to make the effort necessary to accomplish it. The choice is up to the individual concerned.

A further point might also be made about the possibility of cramming. Several organizations run special weekend courses for the written parts of examinations, which seem to work very well indeed. Candidates for the written part of the commercial or instrument test attend study sessions from Friday through Sunday. On Monday they take the examination. The passing rate is high, but critics suggest that retention of the learned material may not be as good with these students as with those who have studied the hard way.

Because the regulations were changed on November 1, 1974, it is important to know what the new requirements are. Perhaps the most important change is the requirement that proficiency be demonstrated in all instrument approaches—VOR, ADF, and ILS. ADF and ILS approaches may be simulated. Candidates are still required to have either a commercial pilot certificate (for which an instrument rating became a requirement) or, if a private pilot, at least 200 hours of flight time. This must include not less than 100 hours as pilot-in-command, of which 50 must be in cross-country flight in the category of aircraft for which the instrument rating is sought. Similarly, 40 hours of instrument time are required, of which not more than 20 hours may be in a simulator. (Details are spelled out in FAR Part 61.65 Instrument Rating requirements.)

There is quite a bit of book learning in addition to having to learn to fly an airplane all over again. I say all over again because the ability to fly by reference to instruments does not seem to be a natural ability. It has to be learned; each instrument, although capable of providing its own data, must also have this data coordinated with that of other instruments for the entire picture to be understood. The worst difficulties are at the beginning; once you develop some confidence (keeping the wings level is a good key here), you will find that the growing excellence of your technique will transpose itself into your normal flying—that is, to your VFR flying.

PHYSIOLOGICAL ASPECTS OF FLYING

The reason that instrument flying is difficult at first is physiological. If you blindfolded yourself and tried to walk across a strange room, how well would you succeed? Most of us would quickly trip over some obstacle and fall on our faces. In the cockpit, however, there are instruments that can tell us—once we have learned to read them—what we are actually doing. But here some confusion occurs. For although we cannot see out of the windows, our other sensory mechanisms continue to play tricks with our minds; we tend to believe what these senses tell us rather than what we might read for ourselves from the instruments.

Our sense of balance is the first source of confusion. It may tell us the airplane is turning, when in fact it is not, or that the machine is climbing or descending, when in fact it is straight and level. For this reason the would-be instrument pilot has to become aware of the circumstances in which he is likely to be confused and to learn to mistrust the signals of his sensory mechanisms and to rely on the aircraft instruments.

The study of the physiological factors which relate to instrument flight is a fascinating subject for most people. First we study the effects of atmospheric pressure on the human system, how changes in pressure can cause the body to function either more or less efficiently, and what we may do to mitigate any adverse changes. We learn about the effects of reduced atmospheric pressure and how these may in some instances cause discomfort and in others endanger our very being. We learn the effects upon the body if you deprive it of its required oxygen. In short, we learn the consequences to our bodies of being in an environment for which it was not built.

Carbon monoxide, an odorless and colorless gas, is an ever-present danger not only from mechanical sources but also from the smoking of tobacco. Three cigarettes smoked at sea level can raise the body's physiological altitude from sea level to as much as 8,000 feet above sea level. The trouble with carbon monoxide is that it will combine even more readily with hemoglobin—the oxygen-carrying substance in the bloodstream—than oxygen itself. Carbon monoxide actually combines with hemoglobin some 300 times more readily, and once the two are together, their resistance to separation is in the same proportion. There's even a further problem. Regardless of the pressure at which oxygen is being exposed to hemoglobin, the existence of some carbon monoxide reduces the amount of hemoglobin available to carry oxygen and the carbon dioxide into which the oxygen is converted after use by the body.

Alcohol is another poison. The instrument pilot learns what alcohol can do to his body. Culturally, we have many misconceptions about alcohol, such as using it as a stimulant (which it is not) or exercising to help burn up alcohol within the system and mitigate its effects (which exercise won't do). Other ideas are that a person who can hold his liquor on the ground will have no trouble holding it in the air—also not true. Two ounces of alcohol in the system is doubled in its effect on body response at 10,000 feet.

Cold capsules are another drug that affects the system. It should be categorically emphasized that any pilot who gets a cold should remain on the ground until he has recovered. People taking pills have no business at the controls of an aircraft.

In addition there are false sensations that may occur during instrument flight from various maneuvers. This is nothing to be frightened about; understanding the sensation (for example, vertigo, a type of spatial disorientation) and the circumstances in which it can occur will help you deal with the situation.

AERODYNAMIC THEORY

A deeper understanding of aerodynamic theory is required. None of it is very difficult; much is plain common sense and will be useful later on in the efficient use of radio aids for navigation.

You will also be taking your first steps toward instrument flight in a simulator. There are several types of simulators on the market. Some, such as the one shown in a popular airline commercial, come with everything from engine noise to ground fog on short final. Most ordinary ground schools, however, have less sophisticated models—but you can learn from them as well.

The most difficult thing to learn is not just straight and level flight by means of instruments, which is quite easy. What does require effort is learning the amount of lead time required and the exact pressure needed to change direction, either left or right or up and down. While one knows, in theory, that both aileron and rudder are needed to start a turn, the average VFR pilot has a tendency to put in a quite arbitrary amount of "push" to get it going, correcting by feel once it has begun. The instrument pilot on the other hand, must know precisely how much force is required to start a turn.

Minimal forces are what's needed, and this means monitoring motor inputs by means of the instruments. Instrument flying requires the right amounts of both aileron and rudder for a turn. Still more important, the pilot must plan ahead: He should have already worked out in the back of his mind how much turn will be required to complete a given course change and approximately at what state of the turn he should start to roll out of the turn. In other words, the flying of the aircraft must be second nature, leaving the mind free to think ahead to the tuning of radio aids, estimates and position reporting, etc.

Exactly the same applies to ascending and descending. For example, at 2,700 rpm it will climb at 500 fpm up to 8,000 feet. At 2,100 rpm it will descend at 500 fpm. The secret here is to know the numbers for the particular airplane you are flying.

The simulator is where you take your first steps in acquiring this proficiency. Perhaps the best part about using simulators is that all but the most basic use an attachment that marks out the course you've been flying on a chart beside the machine. You can practice precision maneuvers and then stop the machine at any point to see if you have been doing them correctly.

Getting out of the cockpit simply cannot be done while flying an airplane, and it is this factor that makes a simulator so useful in the early stages of instrument flight. As you get better, you'll start flying by instruments for real, but until you have a fair knowledge of the basics involved in actual precision piloting, you'll save yourself money by using the simulator. The object here is to train you in the correct method of reading your instruments and to enable you to learn to use the controls in response to that information. At the beginning you will start with a full panel. As you progress, you learn to fly with the emergency or partial panel.

The full panel includes artificial horizon, directional gyro, ASI, sensitive altimeter, turn and bank, tachometer, magnetic compass, and clock. The partial panel omits the artificial horizon and the DG. FAA regulations actually stipulate that a would-be pilot demonstrate his ability to control the airplane using only the emergency panel, even though you are not permitted to file an instrument flight plan without having a full panel in working order.

Apart from actual systems failure, both the DG and the artificial horizon can fail temporarily after they've been pushed beyond certain degrees of bank and/or pitch. In such maneuvers the remaining panel is unaffected. Another cause of actual malfunction could be the failure of the vacuum system; therefore the turn and bank should be electrically operated in any aircraft intended for instrument work in case of such a failure. Ideally, the panel should be divided so that if you suffer vacuum failure you have electrical backup, and vice versa, a point mentioned earlier.

You'll probably make your first instrument cross-country flight in the simulator. By now you should be familiar with all the basic maneuvers and should have learned how the artificial horizon is used to indicate nose attitude, registering almost instantly any change by movement of the airplane symbol at center in relation to the horizon bar. You will soon get to think of viewing in your mind's eye, your own aircraft in back of the one in the instrument. But, like almost every newcomer to instrument flight, you will find your attention dwelling on one instrument too long. Your instructor will show you how to make your instrument scan more efficient, and you just have to keep plugging away at it until it becomes second nature.

The next point that is usually covered is changing airspeed while maintaining a constant altitude. This is an important exercise and could usefully be taught at the private-pilot level. The skill lies in coordinating the controls.

We already know that an airplane may be flown without loss of altitude at a high angle of attack, but few of us ever bother to fly close to the limits of either slow or fast. We're content to push the throttle all the way forward for takeoff, reduce to cruise at altitude, and pull back the throttle to come down.

Perhaps the most important thing to remember here is to realize that at low speeds any attempt to reduce the rate of descent by lifting the nose has a good chance of ending unpleasantly, so we learn to control that rate of descent with the throttle. It sounds very commonsensical, but you will be surprised how easily common sense can sometimes disappear in an initially strange environment. And as you progress with the full panel, you will also start learning to do it without the artificial horizon.

The first time you get into an airplane for real, you may find that your memory—if this was the type of airplane you'd done much of your flying in—starts to play tricks. Your ears will recommend a certain rpm range as being correct, and you may try to get back to your former VFR habits. You will discover yourself making control movements in advance of the instruments, especially in air that is

243

less than smooth. Provided that you know of this in advance and that you psych yourself ahead of time, making a firm commitment to avoid any control movement not dictated by the instruments, you will speedily transfer those skills you obtained in the simulator to the cockpit of the real airplane.

Before you get carried away, you must be reminded that you won't have all that much time to enjoy it—not at the beginning. Rather unfairly, you'll be expected to devote the spare time and energy you have released to other work, such as navigation and communication. If you don't have your own aircraft when you are doing your instrument program here's a useful tip, especially if you are unfamiliar with the aircraft type. When you have the chance, sit in the airplane while it's on the ground and learn where all the instruments and switches are so that their positioning becomes second nature to you. This should be done with every subsequent aircraft you will fly. You can even go so far as to close your eyes and see whether they fall to your hand easily. Remember one point, however, whenever you reach for a switch when actually flying: ALWAYS check visually that it is the switch you want and that it is not already in the position that you want it. There are pilots who at one time or another have leaned the mixture instead of "leaning" the prop! But if you make a visual check first, it is unlikely that you'll go wrong. Fortunately, as the design of cockpits improves, the chances for this type of error are being reduced.

The next stage is learning why instruments actually work, how reliable they are, and if they have any peculiarities that need to be considered. Although this information was sketched in for the private-pilot certificate, the instrument pilot—who relies on instruments as his sole means of data—has to know them better. In particular, he must know the principles of the gyroscope, as this provides the basis of the fundamental information provided by the artificial horizon, DG, and so on.

Let's take a look at the modern turn coordinator, for example. It's a much simpler instrument to read than the old-fashioned needle and ball. Keeping the wings level is a much easier chore to monitor, and although the device won't give you pitch information, it is nevertheless invaluable as a backup to the artificial horizon. The turn part of the instrument tells you the rate at which your own airplane is turning about its vertical axis in the number of degrees per second. It also tells you about bank, but for a better understanding you should know the relationship that exists between airspeed, rate of turn, and the angle of bank involved—you should have covered this again when you were dealing with the aerodynamics involved in instrument flight.

The ball part of the instrument is an inclinometer, or indeed, just like the bubble in a carpenter's spirit level. Early bank indicators were just u-shaped spirit levels. If the forces are in balance during the turn, the ball stays in the center. In the skid—rate of turn too great for angle of bank—centrifugal force pushes the ball to the outside of the turn. In a slip—rate of turn is too slow for the angle of bank—the absence of centrifugal force allows the ball to move to the inside of the turn. You may recall being told to "step on the ball" to coordinate your flight path by applying rudder as indicated by the ball.

The instrument pilot meets another old friend in the form of the magnetic compass. Here again it is necessary to have a deeper understanding of how it works. When you're working from an emergency panel, the magnetic compass is your only aid in determining direction. Understanding compass errors and how they occur helps keep you on course.

Each instrument flight improves your scanning ability, although you may not be aware of it. But one day you suddenly become conscious that you're not having to work very hard at keeping needles in their proper place, and somehow everything seems to be all right. This psychological boost seems to occur for most people just when it's needed, at a point when you think you never will be able to get everything together.

Having mastered elementary flight by instruments, including steep turns, stalls, and unusual attitudes, you now add to your repertoire of navigational techniques the use of radio and certain procedural flight patterns. Although most schools concentrate on the use of the omnirange system as a basic navigational facility, you are also required to know how to use nondirectional beacons.

AIR TRAFFIC CONTROL

Before examining in detail how you go about making an instrument approach and landing, a word first about working with Air Traffic Control. Many VFR pilots are intimidated at the idea of having to work with ATC, whose jurisdiction is slowly spreading to lower altitudes across the country. Not surprisingly, most new instrument pilots get unnecessarily nervous when communicating with ATC. The first bout of nerves seems to come with copying clearances, when occasionally the delivery seems rather fast. But there's a trick or two to cope with that. Before you file an instrument flight plan, it's useful to find out what are the preferred routings, as ATC still shies away from using routes that they may have reserved for some special objective or never have worked or heard of before and from dealing with aircraft equipped with RNAV systems. If you check the preferred route for your own journey, you can fit into their computer system very easily; the chances are that you'll simply be told that you are "cleared as filed," given a cruise altitude and a transponder code to set and squawk, and told the departure control frequency.

Most instrument pilots devise their own shorthand to deal with clearances, but the less complicated your own system, the better you'll understand it all. If you analyze what is actually being said, you can simplify very easily. First, you are cleared from one point to another. The next item is what routes you will follow to get there—this is the airway or airways you'll stick with. Occasionally you may come to an aerial crossroads and must change direction, and make your way off-airways to a different route. Having established your route, you'll be given an altitude, or possibly two or three. You could be told to climb to and maintain 3,000 feet for vectors to a given intersection, where you will then climb to your cruise altitude. But even that isn't too difficult. Finally, you're told what transponder code you've been allotted and what frequency you should use next. It really is as simple as that.

Another reason some pilots still have hassles in writing down the information, is that many of them get too involved in trying to interpret the clearance while it is still coming through. First write down the clearance accurately, then start worrying after you've read it back. A quick check will let you know whether it is going to take you unacceptably out of your way. And if you are cleared— say from mid New Jersey to Boston—over a large part of the Atlantic Ocean, say you won't accept it.

Finally, have a good look at your chart when you file and note the names of the VORs and intersections that just might be included in your clearance. You can avoid considerable confusion and help yourself if you are able to interpret correctly what the man is actually saying. The last point is to have a boom mike so that one hand doesn't have to deal with the mike and the pencil at the same time.

Copying clearances, like almost everything else in flying, really comes down to common sense. And if you don't like the clearance you've been given, refuse it—but be discrete, for it will mean a delay. Even a delay is preferable to finding yourself miles out over the ocean or fighting with ice in a single-engine airplane on a cold, moonless winter's night.

246

DESCENT AND LANDING

The instrument approach and landing, like getting ready for a normal visual landing, is best achieved when you are at least one step ahead of your airplane. If you monitor the Approach frequency (in exactly the same way as you monitor the destination tower frequency on VFR flights), you'll have a good idea of what information you are likely to get and how soon you may expect it. Most smaller airports have only one or two procedures, but the larger ones have more, sometimes many more. If an ATIS is provided on the VOR or ILS, you can get details quite a long way out of what's in use. If there's no ATIS, no Approach Control, and not even a tower, you could ask Center if you can leave the frequency for a moment and call the nearest FSS to your destination for winds and weather. Once you're nearly ready, clean up the cockpit and stand by with your approach chart. Now make yourself believe that you will have to execute a missed approach as a practice exercise.

There is a method to this apparent madness. If you tell yourself you are going to have to make a missed approach, you will be most unlikely to extend your descent beyond safety. The secret of successful instrument landings is to be thoroughly unwilling to push your luck beyond what is specified on the chart. Most weather-related incidents on instruments are caused by pilots who duck under the approach path because they are convinced that the runway is truly visible only a few feet lower.

So to begin with, make sure you know the missed approach procedure. Now all instrument approaches proceed step by step. Get an overview of headings, altitudes, distances, and times involved. Check the navigational aids available, and if it looks like being odds-on for that missed approach and you have the additional equipment, get set up radiowise ahead of time. If your approach is a precision approach, write the Decision Height (DH) you will use and the time allowed on a piece of paper and put it where you can see it; if nonprecision, then write down the Minimum Descent Altitude (MDA) to be employed. Finally, think how the airport is going to appear once you break out—where the runways will be, possible power lines or smokestacks, and any hills in the vicinity.

All approaches are either precision or nonprecision. The most usual of all precision approaches is the ILS, in which the glide slope information is transmitted from the ground to the cockpit. Precision Approach Radar (PAR), the other precision system, is used extensively outside of the United States but is rare here; our radar just doesn't appear to be up to it. When PAR is working, the landing aircraft's profile in respect to the glide slope is shown on the radarscope, and the controller tells the pilot what corrections if any are needed.

The object of the ILS system is to permit a pilot to locate a particular runway and to descend to it in conditions of poor visibility. Developed from the Lorenz system—an early German radio aid— the contemporary ILS evolved from the Standard Beam Approach (SBA), in which audio signals told the pilot whether he was lined up on the runway or to its left or right. In addition, an outer-marker and inner-marker signal told the pilot he should have descended to 600 feet and 100 feet at each point. There was no glide-slope control, however, and the system was operated on high frequencies.

The modern ILS is a considerable improvement over SBA in that basic data is presented visually, with an audio assist. The ideal glide path is displayed continuously so that a pilot can correct at any time. Accuracy is good, as very slight flight deviations produce large indications of error on the instruments.

For clarity, Figure 11.1 has exaggerated the limits of the glide slope, the aircraft being shown to fly between the upper and lower parts of the sandwich or signals. From the far end of the runway a localizer signal is beamed in two patterns—the right-hand sector

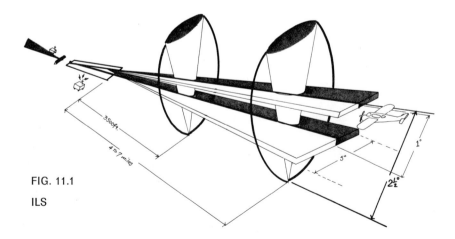

FIG. 11.1

ILS

transmits a 150-Hz signal (blue area—dark in Figure 11.1) and the left-hand sector a 90-Hz signal (yellow area—light in Figure 11.1). At the center is an overlap of the two signals that produces a 5-degree beam (4 degrees at runways more than 10,000 feet long), which at ten miles from the runway is almost a mile wide. The localizer signal is designed to provide its on-course signal to a minimum of twenty-five miles from the runway at a minimum altitude of 2,000 feet. It may be received as much as seventy-five miles away if an aircraft is at a sufficiently high altitude.

A two- or three-letter identifier code is also transmitted to allow the pilot (or navigator) to confirm that the aircraft is flying by the correct signal. The navigation signals from the localizer are fed to the omni head, which, because of the frequency selected, increases its sensitivity from 2½ degrees per one-dot deflection to 2½ degrees per four-dot deflection. Thus, needle deflection to the right (in the yellow sector) means the pilot should fly right to get back to the center; needle deflection to the left (in the blue sector) means the pilot should fly left to get back to center. When flying a back-course to an ILS (approaching the reciprocal runway) the pilot must remember to fly away from the needle.

Glide-Slope Transmitter

The glide-slope transmitter is essentially a localizer on its side. It also transmits two beam patterns, which are up and down rather than left and right. Its frequencies, which are automatically selected by the radio, are in the 328.6-to-335.4-MHz band. The transmitter is usually placed to one side of the runway, not far from the near end.

Unlike the localizer, the glide-slope signals form a beam of only about 1 degree, which is adjusted to give a descent path of approximately 2 to 4 degrees, with 3 degrees being the norm. The glide-slope transmitter does not provide backcourse information. Its width at ten miles is much narrower than the localizer beam, being about 920 feet. This means that the moment your glide-slope needle starts to pick up the signal you begin to descend.

Glide-slope transmitters tend to transmit false signals or echoes of the proper signal, usually at an angle of some 12 degrees to the horizon. Check your altimeter.

Marker Beacons

The outer and middle markers provide the pilot with information about how far along the runway center line he has progressed.

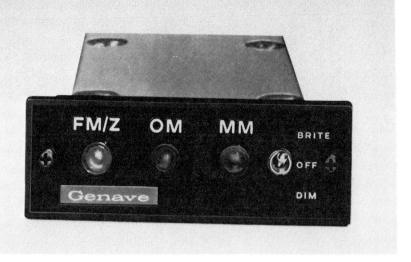

FIG. 11.2

Genave Delta/303 marker beacon receiver

These transmissions are beamed upward in a fan shape (hence they are sometimes called fan markers) and are automatically received by the marker-beacon receiver on 75 MHz. The visual signal is modulated at the outer marker by a 400-Hz audio tone, sending dashes at about two per second. At the Middle Marker the audio is 1,300 Hz and sends alternate dots and dashes. The Outer Marker may be anywhere from four to seven miles from the airport and is generally at that point at which the glide-slope signal intersects the minimum holding altitude. The Middle Marker is usually about 3,500 feet from the runway threshold. Inner Markers are found at Category II airfields and at certain "hard-to-land-at" airports, such as those surrounded by mountains.

Compass Locators

Compass locators are simply low-power nondirectional beacons that, when used with an ILS front course, is co-sited with the outer and middle markers—shown as LOM and LMM on approach charts. Your ADF can be tuned to these beacons to provide additional course information. One point to note is that a number of LMM have been decommissioned, so check your charts and the AIM.

Approach Lighting Systems

Not shown in Figure 11.1, although an integral part of the ILS at busy airports, are high-intensity approach lights and the Visual

250

Approach Slope Indicator (VASI) system. The high-intensity lights provide a brilliant blue-white burst of light that appears like a moving ball traveling in the direction of the runway.

The VASI system provides the same information as the glide-slope receiver but by means of colored lights. Each two sets projects a split beam, white in the upper half and red in the lower. If you see both a red and a white, you are on course. If you see white only, you are too high; if red only, too low. In smog and poor visibility approach lights are difficult to read unless you're close.

An ILS is what is called a forcing procedure—the further you get into it, the more accurate you must be. The trick of flying an accurate ILS is to get the gauges wired at the earliest possible moment. Fix that rate of descent on the button, glue the localizer needle straight down the middle, and suspend the glide slope bar right where it should be—dead center.

An ILS approach may only be flown if the localizer, glide slope, and outer and middle markers are working. Only when they're all working are you permitted to come down to the ILS minimum. If you don't have all the equipment in your airplane—or if the ground gear is partially inoperative—the minimums are higher. The ILS ground gear is often augmented by an Approach Lighting System, which appears from the cockpit as that ball of light leading you to the runway. The Approach Plate will tell you what minima apply when any part of the ILS is malfunctioning.

Localizer frequencies are always odd tenths—109.1, 109.3, and so on—and once you have tuned the appropriate frequency, the glide slope receiver, whose frequency works exclusively with the localizer frequency, is automatically tuned in. Once tuned to the localizer, the OBS is switched out of the system.

One of the nicest things about flying an ILS is that usually you'll be vectored to the localizer. This really makes the work load bearable because you won't have to bother with procedure turns. If you look at an approach plate, you'll notice that usually several approach routes are marked to the outer marker. The arrows (on the Jeppesen charts) give the course, with the distance and minimum safe altitude you may fly. Some routes that are only a few degrees away from your final approach course do not require a procedure turn anyway; these are marked NoPT.

Your clearance is now given to you—for example, "Goshawk 20400 is cleared from the Washington VOR direct to the Washington compass locator, descend to 3,000 feet, cleared for ILS runway 24 approach." You acknowledge the clearance and start your descent, tracking the ILS beam while inbound to the outer marker. If you have two VOR receivers, tune the first to the localizer—check its identity first—and the second to the Washington (in this example) VOR. As you close on the outer marker, the needle in the first set will start to quiver, which warns you to standby for your outbound turn.

Careful planning will ensure that you are at procedure turn altitude by the time you are inbound, and you will then be able to make a comfortable and controlled descent. (Immediately after you've finished a procedure turn, make a careful check of what heading is needed to keep you on the localizer. If you haven't already done so, make sure you're at the right altitude to pick up the glide slope at the outer marker.)

The glide slope now starts to move, which tells you that you're almost at the marker. Your airspeed should now be appropriate for a moderate rate of descent and landing. If ATC wants you to move faster and you feel this may compromise the task in hand, simply tell them you can't comply and let them sort it out. Your job is to fly the airplane down safely; theirs is to sort out and control the traffic.

When the glide-slope needle is centered, immediately begin your descent, keeping both needles steady. This is permissible, even though you may not have passed the outer marker. Check your altimeter, though. Once on course and on glide slope, stand by to let the tower know where you are—for example, "Goshawk 20400, marker inbound." The tower will now come back with permission to land, as they've already been informed of your impending arrival by phone from Approach.

As in localizer tracking, the glide slope is maintained by small alterations of power. If the needle goes down, fly down; if it goes up, fly up. Use discrete amounts of power. Correct for course with discrete amounts of rudder. The changes required will be very slight, therefore a light and delicate touch is needed.

It is at Decision Height that you must decide whether you are going to continue with the landing or execute the missed approach you've been preparing for. If the runway is visible and you're in a reasonable position to continue the landing, go ahead. If it isn't, then you'd better get right out of there. If you pass the decision height and then the runway vanishes, you must make an immediate missed approach. Apply takeoff power, bring the airplane to the pitch-up attitude normally used for initial climb, and check that you have a positive rate of climb on your VSI before cleaning up the aircraft (retracting undercarriage and flaps up by stages).

Assuming all goes well and that you continue on down from DH you now keep up the instrument scan, adding outside visual cues to it. If the runway does disappear, you'll still have sufficient instrument information to be able to abort the landing. Most pros continue the glide slope angle right down to flaring before touchdown. The signal disappears about one hundred feet from the runway.

Unfortunately, current glide slope equipment does have its off days; all too frequently this means that you'll have to fly a simple localizer approach. This is a nonprecision approach, which means higher minimums. Instead of a Decision Height, you will have a Minimum Descent Altitude and must rely on timing to determine your missed approach point. The procedure is similar to flying a VOR approach, except that with the localizer you have much greater accuracy in lining up the runway. As with the VOR approach, you determine at what point you'll begin the descent, then reach your MDA and finally decide whether you'll conduct the missed approach.

The principal difference between flying an ILS and a localizer approach is that with the latter, assuming there is no ground obstruction, you plan to arrive at the minimum descent altitude ahead of time. This allows you to look for the elusive runway just a bit longer. Whether on a precision approach or a nonprecision approach, always be conscious of what you are doing and don't let your attention wander.

We've looked at but a few of the aspects of flight with instruments. When the weather is poor, and all the VFR people are grounded, that instrument rating more than pays for itself. It gets you flying, and if your business is important, it permits you to get there—just like an airliner. For reasons of space this look at the Instrument Rating has been brief—if you're interested in pursuing it further, two publications can be of great assistance. First is Richard L. Taylor's *Instrument Flying*, Macmillan, $6.95. The other is *Flying* Magazine's *Guide for Instrument Flying*, priced at $1.50. Finally, the Jeppesen home study course, for about $30, is good value for the money.

THE PROCEDURE TURN

Whether you're flying a precision or nonprecision approach, there's one useful maneuver you should learn—the procedure turn and its variations. This turn is not used in all approaches, but it is perhaps the handiest of weapons in the repertoire of the instrument pilot. Procedure turns are used for reversing course. They work by blending time with course changes. The simplest of all is the sixty-second

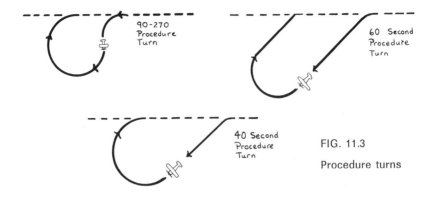

FIG. 11.3

Procedure turns

turn, which requires at least three minutes from the time you leave your outbound course, until you're back on the reciprocal. The purpose of the procedure turn is to get your aircraft lined up for the approach or to park it in the air in a holding pattern. Having left your departure point, you have successfully navigated almost to your destination. At the fix, you turn on to the reciprocal course to that which would bring you to the field. In order to get on to the course for the field you employ a procedure turn. (It should be mentioned that there are fields where the procedure turn has no part in the instrument approach.)

Your sixty-second procedure turn begins as follows: First check the chart to see which way the turn should be made, then start your standard-rate (3 degrees per second) 45-degree turn away from your outbound course and fly for one minute on the new heading. At the end of this minute make another standard rate turn *in the opposite direction to the first turn* for a full 180 degrees. You have now about thirty seconds before you'll pick up the inbound course to the field, and as soon as the needle starts to flicker you can lead your final turn to your inbound and be all set for landing.

The sixty-second procedure turn is best used when there are only moderate winds, for with strong winds you are subject to wind drift and being blown off course. This turn is also very handy if you need to lose altitude on the way to the field.

Two other maneuvers are the forty-second procedure turn and the 90-270 turn. The former begins in the same way as the sixty-second turn, as does the timing. This turn is useful when there's more than moderate winds, as you can adjust your turn to correct for wind drift. Suppose your procedure turn takes you downwind. You subtract one second for each degree of crab you need on your outbound course. Now as you turn back into wind, you add one second for each degree, thus changing the radius to compensate for drift.

The 90-270 turn is a very speedy turn, handy in minimum weather conditions, as you don't have to worry about any math. It's also good in high winds, as you will turn in a minimum of time with a minimum of displacement. The secret is to make all turns at standard rate. Having completed the full 90-degree turn, you immediately roll back into a standard-rate turn in the opposite direction for a full 270 degrees. Depending on wind velocity, the needles will start flickering around 40 degrees from your inbound heading. This turn is full of potential and may be altered to an 80-260 turn. Your key is in making all turns accurate.

From the fix on your outbound course, you'll start timing how far you have to go. This depends in part on how much altitude you still have to lose and on how much time you think you'll need to get readied for the final approach. Don't forget there's a ten-mile limit anyway, so as you approach the fix it's sensible to get your airspeed down to a reasonable level, unless there are airliners occupying the space around you.

PART VI

12

The Wonderful World of STOL

Did you ever think how wonderful it would be to walk out into your backyard, load up your airplane with luggage and family or friends, and then, after a brief takeoff roll, head out for the wide-open spaces? Toward the end of World War II the seers confidently predicted a small helicopter in everyone's garage or carport by about 1948. But it was soon discovered that helicopters just couldn't be made inexpensively and that one had to keep replacing their working parts at a prodigious rate, which is quite costly.

At about this time people began to reexamine short takeoff and landing (STOL) aircraft. What is important about STOL from a pilot's point of view is its safe flying characteristics during slow flight. You can rotate any airplane early on takeoff, within ground effect, and let it build up speed before you climb, and most pilots can do this once they get to know their own airplane. But this isn't real STOL, it's merely pilot technique.

There are two ways in which you can achieve STOL performance. You either design your airplane so that it has relatively high power and a low wing loading because of large wing area, or you make use of one or several mechanical devices to make your wing more efficient throughout all aspects of flight. For example, a Piper Cub achieves its performance through a relatively low wing loading. The French Rallye Minerva and the Helio Courier, on the other hand, make use of automatic slats in the leading edge of the wing (which pop out at slow speed) as well as special flaps and ailerons. Peter Garrison's *Melmoth*, an original two-seat design of his own, uses a number of sophisticated techniques—able-slotted Fowler flaps, spoilers, stall fences, amongst others—and while not

The Fairey Rotodyne, years ahead of its time could have produced city center to city center travel more than ten years ago. But an old-fashioned and out-of-cash British government scrapped it.

De Havilland Buffalo is one of the all-time great STOL aircraft. Produced by the Canadian company it is in service with military forces in many parts of the world, while a civil version does duty in back country areas where airstrips are short.

Seaspeed's SRN-4 hovercraft "The Princess Anne" plies regular cross-channel shuttles daily between Dover, England, and Calais and Bologne in France.

A small section of metal is attached to the leading edge of the wing in order to provide for standard stalling characteristics on both wings. The metal breaks up the flow of air over the leading edge at slow speeds, actually provoking the stall, but because of its shape and the fact that it is fitted to both wings, the stall occurs at an exact speed, and any tendency for one wing to drop in the stall is partially, if not fully, mitigated.

strictly a STOL airplane, will provide STOL performance plus high cruise (200 mph) when not loaded with fuel for transatlantic travel.

There is, actually, a third way to achieve the sort of lift needed for STOL, although it isn't mentioned very often—the Channel-Wing, a unique design using twin engines, in which the air is taken in over the surface of semicircular channels by the propellers, thus providing lift in addition to thrust. (See Figure 12.1.)

These characteristics provide an airplane with the ability to get off the ground in a relatively shorter space than for either a fixed-wing aircraft or autogyro. You need to achieve a lift greater than the aircraft's actual weight at a relatively slow speed. You also need rapid acceleration to reduce roll drag on the ground. Finally, you want good slow-speed climb performance to lift the plane over obstacles as soon as it is airborne.

Although the actual technology of STOL is quite recent, the devices the engineers use to achieve STOL performance have been around a long time. The automatic leading-edge slats used by the Rallye series (and by the American Helio Courier and Stallion aircraft) were first used by the British Handley Page company in the 1920s. Droop ailerons (found on the Robertson company's modified STOL aircraft) were used on certain World War II aircraft. (The ailerons droop when certain positions of flap are used.) Slots (holes in the forward edge of the wing to achieve better airflow at low speeds) were also used during World War II. Leading-edge slats, slotted flaps, and area-increasing flaps are just part of the stock in trade of the STOL design engineer. In addition, stall strips, which are fixed to the wing's leading edge, make the wing stall at a particular speed; stall fences prevent the stall from

FIG. 12.1

The Custer Channel-Wing

getting outboard of the wing and reducing the effectiveness of the ailerons at low speeds. On top of the wing there may also be spoilers, which provide roll behavior when the airspeed is too low for the ailerons to work effectively and which can also be used to reduce lift and increase drag. Finally, the use of two or more engines utilizes the propeller slipstream as an additional factor to augment lift on larger airplanes. All engines and propellers are interconnected by cross-shafting, which in the event of an engine failure allows the remaining engine or engines to power the failed engine's propeller.

However, getting the plane airborne in a short space is not the whole story; a STOL aircraft must also be able to land in a short space. Slow speed in touchdown is needed to reduce the landing roll. The flaps, which are used to provide extra lift on takeoff, are also used to slow the plane down. Lowered even more than in the takeoff position, they still provide lift but with increased drag. (See Figure 12.2.) Even the big jets make use of this characteristic, and if you've ever sat in a window seat at or behind the wing, you will have noticed how the flaps curl right down when the plane is put on final approach. Obviously, a lot of power is needed to keep the machine flying with all that drag.

Birds use the same idea. If you have ever watched a sparrow or pigeon coming in to land, you will have seen that just prior to touchdown the trailing edges of the wing feathers are brought almost through 90° and the tail feathers are brought almost forward. In like manner the STOL aircraft lowers its wing flaps to increase drag and lower the touchdown speed. Some planes use a "lift dump" switch, which automatically retracts the flaps at touchdown

FIG. 12.2

The Fowler type flap seen extended here on this SOCATA Rallye provides considerably increased wing area with relatively light drag. The Fowler flap is one of several lift improvers available to wing designers and is found on many STOL-type aircraft.

263

to cut all lift and may also use wingtop spoilers to break the flow of air over the wings. Some aircraft also make use of reversible pitch propellers, which can be used in the air to provide considerable drag with a high rate of descent. On landing, the pitch is fully reversed to provide additional braking.

Instead of automatic slats that pop out when the speed is low, another way to achieve the lift needed is to recontour the leading edge of the wing, as the Robertson company does on their many modifications of standard aircraft. The Cessna company has also introduced this cuff type of leading edge on two or three of their airplanes. Most jet aircraft make use of forward slats, stall fences, and slotted flaps so that their wings, efficient at high cruising speeds, can provide adequate lift when they slow down.

You may wonder why, if STOL is as easy as described here, more manufacturers don't make their planes STOL to begin with. The standard wing, you may think, cannot be that efficient and must be something of a compromise, and it is. The reason is partly the result of the human inertia of design engineers, and partly the dictate of the accountants, whose rule over modern corporations stifles, or tends to stifle, all but the simplest and least costly innovation. Finally, there is the question of pilot technique: It takes about ten hours to become modestly proficient in a STOL airplane and about thirty hours to become good. According to market researchers, pilots are not prepared to devote that much time to learning the new techniques, though as a pilot I don't agree. Most private pilots are keen to add other skills after obtaining their licenses, such as the instrument rating. And the STOL checkout can usually be accomplished in about 10 hours or less.

Still, more and more pilots are discovering that they have a requirement for STOL or near-STOL performance. It is not really so much a question of safety, although safety is an extremely important factor. (For example, the risk of a serious landing accident at 130 mph is four times greater than at 65 mph). STOL not only increases the safety aspect of airplanes but also increases the airplanes' entire performance, enabling pilots to utilize their STOL aircraft to a much greater degree than a standard machine.

Faced with declining importance over the next twenty years (according to predictions of the FAA), a result of the increase in numbers and use of general aviation aircraft, the airlines are looking for better ways to serve the public. As of January 1, 1972, general aviation (as private flying is more usually referred to) accounted for some 98 percent of all U.S. civil aircraft, 135,000 in all. Of these, half were single-engine, four-seat (or more) machines. General aviation also accounts for 96 percent of the nation's pilots, who in the year past flew some 30 million hours. And in 1969, for example, general aviation carried an estimated 177 million Americans, compared with the 166 million carried by the commercial airlines. Actual figures quoted by the FAA indicate that more than 90 percent of all passenger air traffic across the nation will be handled by private aircraft by 1990.

One way the airlines could fight this trend would be through the introduction of STOL. By using STOL an airport could, in theory, be almost in the city center, as were the early railroads. Passengers could fly direct from center to center without those long boring journeys to outlying airports and the equally boring waiting to get on the plane, as well as the additional time consumed as the airplanes wait their turn to get off the ground.

However, basic to any implementation of a STOL system, either in the United States or elsewhere, is the location of the airport. Whether the STOLstrip would be at a conventional airport, feeding in passengers from outlying areas, at special landing pads on the tops of buildings for V/STOL operations within the cities, or at specially created STOLports, exurban and metro—all these are basic to any discussion about STOL in commercial use.

Before STOL airplanes start making their appearance around cities and suburbs, the public has to be convinced of their merit and the inherent benefits of permitting these aircraft to take off close to residential areas. Airplanes, albeit infrequently, still manage to crash into buildings from time to time even near the best-regulated airports. But the siting of STOLports is something different. The minimum acreage required for a commercial STOLport is about twenty acres, with just a little more if you want to provide reasonable ramp space for aircraft. Runway length has been placed at 2,000 feet. (The FAA has regulations covering the specifications and performance of STOL transports.)

Quite apart from the problem of noise from commercial STOL aircraft, STOL has yet to be shown to be better than other forms of transportation. In these times feelings run high about the need for environmental protection, and the establishment of what seemingly is yet another noisy and polluting means of transport is a hotly debated question. To be accepted, STOL must be shown to be financially feasible, cheaper to the traveler, and profitable to the operator before it becomes an acceptable alternative. And it must be shown to alleviate the chronic inefficiencies and delays that have become a nearly daily occurrence at almost all our larger airports. Finally, the so-called energy crisis has made a profound difference to management thinking.

This has put STOL in competition not merely with conventional air transport but also with other means of transportation, including hovercraft, hover-rail, railroads, buses and so on. There are many who feel that with increased efficiency in design and maintenance, our railroads can provide a worthwhile alternative to crowded airports and short-haul commuter road traffic. The Japanese have proven high-speed rail transport to be a workable alternative, and in Europe, the Trans-European Express already offers travel times that permit it to compete very effectively with air transportation on short routes and with considerably more comfort.

To compete with a STOL system in the United States, a rail system would need to provide cruise speeds of up to 150 mph that could be reached within five minutes of leaving a station. Cruise speeds of up to 200 mph are certainly possible, and such a system could offer lower fares, greater fuel efficiency, and less pollution than the airlines' system. In terms of profitability it could well be the long-term winner. The snag is that it would cost considerably more to put into operation than a STOL system.

Air-cushioned vehicles (ACV) have also been considered a possible alternative to high-speed railroads. Extensive experimentation has been carried out in France, and evidently this system could be implemented almost at once. The ACV's main drawback is noise, though the actual equipment itself is quite cheap. What might be expensive would be a suitable track for an ACV permanent way, although this might be solved by building on stilts, using existing roads for some cross-country portions.

One possibility that is being experimented with is the use of an air-cushioned landing gear for STOL aircraft. This would permit both conventional land and water operation of STOL airplanes and would obviate the need for proper runways. The system works extremely well and has been evaluated for military application.

Hovercraft seem unlikely to compete with ACVs and STOL planes very directly except where they are used over water. In Britain a hover-ferry service has been operating between that country and France; it is capable of carrying buses, trucks, and cars, as well as some 700 passengers. Cruise speed in the order of 70 mph has been reached in smooth to moderate seas, and the hover-ferry has had a reasonable record of reliability. Hovercraft are, however, rather noisy to people outside, and maintenance of the flexible skirts, which maintain the cushion of air underneath the craft, is still quite costly.

Hovercraft and hydrofoils (which ride upon curved wings in the water) seem quite practical over inland waterways but are not too

FIG. 12.3

Too much noise on takeoff?

usable over open seas. Experience with hydrofoils in exposed waters tends to confirm this.

The Canadians have already designed a 48-seat "quiet" STOL transport, the De Havilland DHC-7. Engine and propeller noise combined produce a low exterior noise level of 95 pnDb (decibels) at 500 feet from its takeoff path. If you want to compare the noise this four-engine aircraft produces with something you know about, an average six-lane freeway produces about 90 pnDb at the same distance. A commercial jet produces about 116 pnDb at the same distance.

The reasoning behind the DHC-7 was that for any downtown STOL system to gain public acceptance, it had to be quiet. The engineers accordingly lowered the propeller-tip speeds to around 0.6 Mach, and the disc loading of the props was kept low by using large four-bladed propellers of more than eleven feet in diameter. A lot of noise is generated by an aircraft propeller, some through its actual speed through the air and some through the thrust it exerts on the air itself. As a result, the designers worked on the propeller-tip shape, the twist distribution, camber, and even the airfoil of the propeller in efforts to ensure quietness. They went even further: The blade is made of hollow aluminum, is filled with a foam filler, and is sheathed in fiberglass.

In order to keep ramp noise to a minimum, the exhaust is ejected upward over the wing, and special work has been done on the air intake to reduce noise. The engines are cross-shafted, and the wing features a number of high-lift devices at the trailing edges and spoilers on top of the wings. The cost of flying has been estimated at a low 2.6¢ per passenger-seat mile on a 100-mile route at a cruise speed of 275 mph.

Another application of STOL is by the crop-duster pilot. With the new chemicals being used, the pilot must face the toxicity of some of the chemicals. Although the FAA runs special programs to alert pilots to the dangers of the chemicals, the manufacturers of crop-dusting airplanes build very carefully to avoid chance leakages. One of the other things some of them do is to utilize STOL to get the airplane and pilot to and from the fields he must work as quickly as possible. An airplane like the Grumman Ag Cat, for example, needs about 750 feet to take off, clearing a 50-foot obstacle at gross weight, carrying a ton and a quarter of chemicals. North American Rockwell's Thrush Commander needs a few yards more to get airborne, and it has a useful load of a ton and a half.

STOLcraft also have application for fire fighting, public health, crop spraying, traffic patrols, weather reporting, construction work, pipeline and cable patroling, air/sea rescue, and so on. More and more of these applications are being filled by vertical takeoff and landing aircraft (V/STOL, or VTOL, as it frequently appears), which brings us to helicopters and gyroplanes.

A helicopter uses a powered rotor for its lift; a gyroplane's (or autogyro's) rotor is usually unpowered or powered only prior to takeoff, relying on forward motion to turn the blades and provide lift.

Acceptance of the helicopter has been slow, as its cost has been unusually high in comparison with conventional air transport. But its growth during the 1960s has burgeoned; today there are now more than 2,000 special helicopter landing pads around the country. And for this year—despite the economic slowdown—it has been estimated that helicopter sales will reach 1,000+ units per year, with the future for small helicopters looking very good indeed.

A helicopter works much like an ordinary airplane, with one or two exceptions. Instead of a normal wing it uses its main rotor to provide lift and also forward (or backward) motion. It also has a tail rotor, needed because the turning force of the main rotor imparts a force (torque) to the fuselage of the helicopter that tends to turn the fuselage in the opposite direction, and it is also used to maintain the heading of the machine.

The flight controls in a helicopter consist of a collective pitch-control device that governs altitude, a throttle that controls rpm, tail-rotor pedals that control heading and also trim in forward flight, and a cyclic pitch stick for steering.

268

The collective pitch control regulates the amount of lift generated by the main rotor; the angle of attack of the main rotor blades
are collectively increased or decreased, hence the name for this
control. However, when the angle of attack is increased, there is a
corresponding increase in drag. To counteract this and to prevent
the slowing of the main rotor, more power must be applied. In the
same way, if the collective pitch is reduced, which lessens the angle
of attack, engine power must also be reduced to prevent overspeed-
ing of the main rotor.

In conventionally powered helicopters the rotor rpm is controlled by a motorcycle-type throttle on the end of the collective
pitch-control device. In helicopters powered by turbine engines,
rotor rpm is maintained by using an engine-governor mechanism
that works automatically.

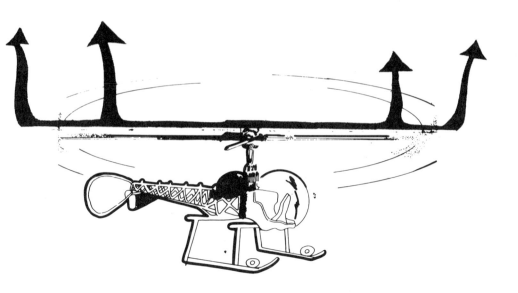

FIG. 12.4

The rotor of a helicopter viewed as a disc revolves on the horizontal and forces
developed act vertically, as lift.

The tail rotor controls the force or torque of the main rotor acting upon the fuselage and is controlled by the pilot by use of "rudder" pedals. To maintain a constant heading in an American helicopter, the pilot increases left-pedal pressure whenever power is increased and right-pedal pressure when power is reduced. In European helicopters it is just the opposite.

The cyclic stick, which looks like the control stick of some conventional airplanes, controls the direction and speed of the horizontal movement of a helicopter. An easy way to visualize this is to think of the main rotor not as two or three blades but as a solid disc. If this disc is horizontally level, the forces it develops will act vertically as lift. If you tilt the disc (with the cyclic stick), part of that lift is converted into horizontal thrust that will move the helicopter in the direction that the disc is tilted. Greater tilt means greater speed.

The cyclic stick, unlike the collective pitch control, increases the angle of attack of each blade only during a portion of each complete rotation. In other words, it causes each blade to produce more lift at one segment of each revolution of the blade and less lift at the opposite segment. The effect is similar to that caused when banking a fixed-wing airplane, when part of the lift from the wing is diverted to turning the aircraft.

Naturally, if you divert some of the lift provided by the main rotor into thrust, you reduce the amount of lift available. To compensate, you increase the collective pitch, which in turn requires more power to maintain engine rpm. And increasing the rpm means that you have to apply more rudder pressure to maintain heading. The interrelationship between the controls in a helicopter sounds highly complicated and is a challenge to the student. But it is not that difficult, although flight time is more expensive learning to master helicopter technique than it is learning to fly a conventional machine.

Igor Sikorsky started building helicopters in Russia as early as 1909. When his second helicopter failed to fly in 1910 he turned to fixed-wing aircraft and flew the world's first four-engine airplane in May 1913. Thomas Edison also designed a rotating-wing aircraft but had other things to do and it never materialized. The man who made possible the modern helicopter was Juan de la Cierva, a Spaniard. It was he who invented the "flapping hinge," which made possible the horizontal and vertical movement of the main rotor blades, providing rotor stability. He fitted this improved rotor into his fourth autogiro, which he successfully flew in January 1923.

After the Bolshevik revolution, Sikorsky came to America and designed a plane, the S-29A (A for America). This plane did quite well financially and helped set Sikorsky's company on its feet. For some years he became best known for the amphibian airliners he designed for Pan American, but he had not given up on helicopters. Just after the outbreak of war in Europe in September 1939, he flew the world's first practical helicopter, the VS 300. Today the name Sikorsky is almost synonymous with the helicopter.

In the autogiro field in Culver, California, Drago Jovanovich unveiled his gyroplane, then called the Jovairin. For a while inventor millionaire Bill Lear tried to spark interest with it, without much success. In the mid-1960s the McCulloch corporation, best known for its oil and real estate interests, took up the idea. After investing more than $3 million in the project, McCulloch finally certified a funny-looking creature with stubby wings, twin boom fuselage, and two seats suspended beneath a large rotor. Behind the small cabin was an engine with a pusher propeller.

The McCulloch J2 gyroplane is almost like a helicopter. It's a two-seater, near-VTOL machine with a cruise speed of about 110 mph and a range of 200 miles. The rotor system is the same as that used by the small Hughes helicopter, which have had some two million hours of service without a failure. Unlike the Hughes machines, the rotor is unpowered and the absence of torque stresses indicate a safety and strength factor that's hardly necessary.

Flying the gyroplane is almost like flying a conventional machine. The starting procedure is standard for the 180-hp Lycoming. There is a mid-panel throttle, connected to a second throttle on the collective pitch control, that is used only during engine run-up. When held down, it engages a belt drive from the motor that spins the rotor shaft. In this down position it flattens the blade pitch from a normal 4 degrees to zero pitch.

For takeoff you line the gyroplane up with the runway and lock the brakes. You then spin the main rotor, which brings down engine rpm until centrifugal force takes over on the blades. Once the blades have reached 550 rpm (there's a dual rotor/engine tachometer), release brakes and the collective and start down the runway. As 30 mph comes up on the ASI, you pull back on the control stick to leave the ground; nudge the stick forward to stay within ground effect until you reach about 60 mph for climb. It takes about 100 feet to break ground on a runway, and about 150 feet in a rough field. Climbing over a 50-foot obstacle, according to the book, takes 600 feet. Best rate of climb is 65 mph for about a 700-fpm climb.

271

In flight the J2 is pretty much like any other light, fixed-wing airplane. It's a bit more maneuverable, especially at slow speeds. With the engine out the plane will descend quietly down at about 900 to 1,000 fpm; in a strong wind with zero speed it'll fly backwards. To land the gyroplane without using the engine, use forward stick to obtain flying speed for the flare. Ground roll is about 50 feet.

The only other gyroplane around is the do-it-yourself Benson, a charming single-seat gyrocopter that buzzes around the skies like a small noisy wasp. It is a rather spartan machine, the pilot protected from the elements only by his clothing. It really is loud; you certainly need some sort earplugs before you take one up.

Gyroplanes still make a lot of sense, as they are much less expensive than helicopters. But at $20,000, the McCulloch J2 is not exactly cheap.

The Scorpion (Rotorway) Aircraft company, based in Arizona, markets a do-it-yourself single- and twin-seat helicopter, for which a cruise speed of around 75 mph and a range of about 160 miles is claimed. The service ceiling is just over 7,000 feet, and it will hover in ground effect up to 5,500 feet. Out of ground effect hover is limited to about 3,500 feet. It sounds like a very nice little ship indeed, but Scorpion doesn't seem to want to tell people about it outside of their advertisements.

Normal helicopters start in price in excess of $25,000 for the simplest two-seaters. The cheapest three-seater is the Hughes, just short of $40,000 at the time of writing; the Enstrom, a particularly nice three-seater, costs a little more. The Enstrom has the edge on the Hughes in range and cruise speed; the Hughes has a slightly better service ceiling and better hover range. Both are about the same in useful load.

Larger turbine-powered helicopters with five seats cost up to $200,000; a Sikorsky Sky Crane will cost about $2.5 million. On the other hand, the Sky Crane will lift rather more than its own empty weight, about two tons to be precise, which makes it a pretty efficient piece of moving equipment.

Another use for STOL concerns floatplanes and amphibians. The reasoning behind the early thinking was that because so much of our planet is covered with water—of the 197 million square miles of surface area, only a mere 57 million square miles is land—the greater part of the surface area could be used for landing and taking off by floatplanes and amphibians. As a result, many of the early airliners were amphibians; indeed, some of the record-breaking aircraft of the late 1920s and early 1930s managed to take top honors in competition despite their floats.

A floatplane aircraft is usually a conventional airplane mounted on top of floats that act as its landing gear in the water. Some floatplanes have wheels within the floats and can be used on land. A floatplane without wheels can usually be landed on a grassy strip but will need a wheeled trolley to get airborne again.

An amphibian is built from the very beginning for use in water. Its fuselage is designed more like the hull of a boat than the hull of a plane and is shaped to withstand the impact of quite heavy seas. It is usually much more solid in appearance than a conventional plane. Most amphibians are provided with retractable gear so that they may land on runways. Some amphibians, such as the Republic Seabee, are so rugged that they may be landed on concrete, wheels up, without serious damage to the hull.

There are two amphibians in production at present: the Thurston Teal, a single-engine, two- or three-seater, and the Lake Buccaneer, also available as the Lake LA-4 (there is also a Lake seaplane, an inexpensive version of the LA-4), a four-seater with one engine. Neither is very fast, the Teal cruising at between 100 and 110 mph and the Lake at up to 150 mph. The Teal needs a stretch of water about 500 yards to takeoff; the Lake amphibians require slightly less. When one considers how many small lakes there are dotted across the country, the usefulness of this type of craft becomes evident.

The most popular amphibian for the private pilot is the Republic Seabee. Unfortunately, it has been out of production for years. It is a large airplane, the cabin being over nine feet long. The seats recline into a full-size double bed. It has a single engine and is usually fitted with wing extensions, as it lacked performance at high altitude lakes and fields. Low cruise speed is about 90 mph and high cruise speed is 115 mph; range is eight hours and five, respectively, with the seventy-five-gallon fuel tank. One item that owners enjoy is the reversible pitch mechanism in the propeller: On land or in the water you can make it taxi in reverse. Good Seabees are expensive.

However, it is in the realm of floatplanes that we find most applications of STOL. The reason is that floats add considerable drag to a standard aircraft's performance; the general rule of thumb is 20 mph deducted from cruise speed and about 20 percent added for takeoff run. (On land the Maule Strata Rocket requires just 600 feet to get airborne over a 50-foot obstacle. On the water, however, the same airplane requires 450 feet. The Piper Super Cub on land requires just 700 feet to get over a 50-foot obstacle; on water it needs just over 1,000 feet.) With STOL modification, however, an ordinary airplane once fitted with floats becomes much more versatile, and it is not too surprising that many so-called standard aircraft that actually have STOL or near-STOL performance already are popular among the floatplane fraternity.

Typical of these are the Cub and the Maule. The Maule is a quite incredible airplane in its standard form. It is old-fashioned in appearance and has been rather cruelly described by its detractors as "the only homebuilt aircraft in regular manufacture." Its owners adore it not merely for its cruise but also for its STOL performance.

The Helio Courier is a popular buy as either a floatplane or as an amphibious floatplane (meaning it will land on land, too). Robertson-modified Cessnas are also coming to the fore in the floatplane field. The Robertson STOL Cessna 206, for example, requires about the same distance on water as the Maule to get up and away. The unmodified version of the same airplane needs nearly 2,200 feet, almost half a mile more.

Basically, the Robertson company modifies an airplane in three ways. First, they recontour the leading edge of the wing. They then install a stall fence between the aileron and the flap. Finally, they droop the ailerons and integrate the aileron system with the flap system. The ailerons still work as ailerons, but when you lower flap, you also lower the ailerons too.

The reason for recontouring the leading edge is to increase the angle of the wing at which it will stall and at the same time to reduce the speed of the airplane at which it will stall. Because of its airfoil design, a wing will stall at a certain positive angle of attack and speed. When it reaches that angle, the air breaks away from the wing, destroying lift, or creating a stalled condition. A wing can stall from either the leading edge or trailing edge.

If it stalls from the leading edge, the stall is quite abrupt, almost violent, and usually a wing will drop. A trailing-edge stall, on the other hand, is relatively gentle. Frequently evidence of a trailing-edge stall will be indicated by a relatively high rate of sink while the airplane is still controllable. The leading edge of the wing is therefore redesigned to permit a greater surface for the air approaching the wing to pass over and is so contoured that the air going over the wing tends to stick to its surface. The Robertson recontoured leading edge is a highly engineered item that optimizes both low- and high-speed performance, a considerable achievement. It is added and joined onto the existing wing, enhancing the overall structural integrity.

The addition of a stall fence contains the stall in the center section of the wing and prevents it from spreading outward to the ailerons. This allows for full control of the airplane by the ailerons right into the stall.

The objective of the drooped ailerons is to provide a full-span flap system along the trailing edge of the wing. This way the entire edge becomes one huge flap, permitting the wing to generate more lift at takeoff and providing extra lift with drag for landing. The Robertson company also uses droop wingtips, which is claimed to decrease drag and vortex flow of air associated with the standard wingtips. The droop tip carries the vortex outboard of the aileron so that at slow speeds undisturbed air can flow across the aileron. The ailerons are also sealed to prevent any air coming through the aileron/wing gap, creating problems on the upper surface.

The Robertson company also installs an automatic trim system in a number of the aircraft they modify. Worked by a cable and spring system, the mechanism automatically trims the airplane when the flaps are lowered. This makes the airplane much easier to fly, as it cuts out a lot of work for the pilot.

Choosing systems . . .

that provide maximum benefit to all.

13

Aviation Today–and Tomorrow

In the field of light aircraft design we may expect to see increasing use made of the Wankel rotary-type engine. Several companies are presently experimenting with this engine, and if some means can be found to increase its efficiency in fuel consumption it is likely that we'll see this engine in service in mass-produced aircraft very soon. Increasing use of lightweight turbines seems also likely in more expensive lightplanes. However, the turbine—because of the more expensive materials required for its construction to deal with high temperatures—is more costly than either piston or Wankel-type engines. As mentioned earlier, the turbine engines are superior to piston engines in smoothness and in the longer time between over-hauls. This need not be true of some smaller turbines, some of which have a very short TBO.

Multibladed shrouded fans seem likely to replace propellers as manufacturers improve the noise characteristics of their products. This will mean a move to "pusher" engines and propellers mounted at the rear of the aircraft rather than the "tractors," or engine and propeller up front, we've become used to, so there should be a spinoff here in a little extra speed for the same fuel consumption. With shrouded fans, noise reductions of around 20 pnDb seem probable because of the shroud and reduced fan tip speeds; cleaner, lower-drag airframes will assist in improving performance per horsepower. These shrouded fans are called Q-fans, and several manufacturers are understood to be making studies of them at the present time.

FIG. 13.1

Cessna's Cardinal fitted with the Wankel rotary-type engine.

The Boeing company has experimented (with Eastern Airlines) with a jet-powered STOL commuter airplane, one which will certainly not be particularly quiet when compared with the Canadian De Havilland propeller-driven DHC-7. The Boeing 737 would use a deflected jet system to get itself airborne in a manner not unlike the Hawker Siddeley Harrier (manufactured in this country by McDonnell-Douglas for the U.S. Marine Corps), a close-support (tactical fighter-bomber) and reconnaissance VTOL machine.

Another possibility is the application of Willard Custer's remarkable channel wing, which has undergone much delay in its application by the aircraft industry and is still not in production. Custer is, not surprisingly, rather bitter, considering the extensive number of hours of test flying his machines have flown. And practical tests together with extensive computer work demonstrate that the application of channels on pure jet transports weighing 100,000 pounds, could reduce their landing speed by no less than 25 percent—a highly considerable factor in terms of passenger safety.

Equipment costs should become less expensive; where the price remains the same, you should get a lot more for your money. Third-

278

generation, solid-state electronics is just one new area in which this will take place. Already one domestic manufacturer is producing a fully digitalized ADF unit for less than $900. Other more exotic items seem likely to find their way into the cockpit of the light-plane. Even simple weather radar for a single-engine airplane, which may also have a capability of warning the pilot of the proximity of other traffic, seems likely within five years. Instead of using the heavy, old-fashioned, rotating mechanical scanner, the solid-state weather radar relies purely upon its circuitry to make its scan. Coverage is of the same type as presently used in light twins —that is, 180 degrees in a forward direction. The unit can be mounted within the leading edge of the aircraft wing.

RNAV computerized equipment is already becoming a reality for the modern four-seater. Once again, the same technical advances that have made possible the inexpensive electronic calculator are responsible. Basically, the equipment permits the pilot to shift electronically a ground beacon to where he wants it to be. RNAV is slowly gaining acceptance by air traffic control and is getting cheaper all the time. We may also expect to see several other interesting developments as the manufacturers seek to diversify the applications of their newly found technology.

Instrument panels may be expected to get a further facelift when NASA studies are applied to the private airplane. There actually is an ideal position for each instrument, and certain instruments may be redesigned so that the information presented is more easily

FIG. 13.2

Cessna's XMC research aircraft on test with multi-bladed shrouded fan, instead of conventional prop.

assimilated and understood. Similarly, impact absorption by the aircraft shell should improve. The Mercedes Benz firm has applied a considerable amount of this type of research to the construction of their own automobiles.

Cheaper lightplanes could become a reality if government moved itself out of the aviation field. The biggest cost in developing a new and inexpensive aircraft is the amount of red tape that goes into it. And there are excellent reasons for the red tape— safety being one of them. A simple, inexpensive airplane (a four-seater for around $6,000) could be designed and produced within a year if some of the industry chiefs took time out to consider where and how it might be sold.

The production of a supersonic business jet aircraft is more likely. Several firms have investigated the problems involved, and provided restrictions are not made on the flight or landing of super-sonics aircraft—they are noisier—we may see them in production before the end of the decade provided the fuel economy problem associated with SSTs is solved. It's not improbable that a small fleet of supersonic business jets might exist before the large manufacturers get around to producing one, for an American supersonic is almost certain, despite the arguments of conservationists. It will be argued that as both the Soviets and the Europeans have produced super-sonic aircraft, in order to compete effectively the United States should have one as well. Counterargument may have less weight the next time around, when the current generation of jet engines will have been replaced by turbo-fans, offering considerably greater economy, quieter operation, and less pollution. This will be an interesting contest to observe if it takes place.

Boundary layer control, although used on military aircraft, still lies ahead. This is a system of controlling lift by removing turbulent air from wings and fuselage by suction.

Steam engines in their present form seem unlikely for aircraft unless there is considerable simplification of the recirculating and condensor systems. Several groups believe this could be the answer to a nearly pollution-free engine. However, with our presently limited technology, it seems unlikely.

Fuel cells, on the other hand, could become a reality quite soon. Instead of filling the gas tanks, we'll clip in two power packs—good for 1,000 miles. Another possibility is the use of liquid hydrogen as the "perfect" fuel. These ideas are all still experimental, but their use may come sooner than we imagine. An interim step may be the use of such additives as alcohol or methanol in regular fuel for increased economy with existing engines.

Although business aviation will be able to adapt itself to the requirements of a ground-based (and needlessly inefficient) air traffic control system, the needs of popular aviation are rather different. There is a real danger that general aviation aircraft may be squeezed out of urban centers in order to permit the airlines a free reign. The danger is that lobbying by the airlines may succeed in persuading the bureaucrats to penalize the private sector of flying, already viewed with some indifference.

But there is another possibility: the belated decision on the part of our legislators to assemble a panel of the best thinkers in the field of transportation to devise a total plan to take care of our transportation needs for the next century. Such a think tank could be speedily assembled. One encouraging point has the appointment of a power or energy czar to supervise and cooordinate the use of all energy resources in the country.

The point is that aviation is actually a useful resource to mankind, as it allows us to compress the time it takes to move from place to place. But it needs to be geared sensibly to overall needs. As Lewis Mumford has pointed out, the primary purpose of transportation is not to increase the amount of physical movement but to increase the possibilities of human association, cooperation, personal communication, and choice. We ignore this idea at our peril. Although it is sensible to continue to improve our technology, this improvement should not be at the expense of the society it is supposed to serve. Whether we voyage to the stars or whether we commute to work, we must elect to choose those systems that provide maximum benefit and are within the reach of most.

Appendix: Aerobatics

Back in the days of those aerial circuses, when the barnstormers used to fly into town, half the fun was the display of precision—and some not so precision—maneuvers with which the pilots whetted the appetites of mortals below to share the skies above. Your three-buck ride did not usually include partaking of such fare, unless you were a very persuasive—and daring—small person.

The art of aerobatics is almost as old as aviation itself. And today, the United States is—until the next round—holder of the World Championship cups with the best male and female pilots, plus the team prize.

Most of the maneuvers we use today were developed out of the necessities of WWI, when aircraft were relatively slow and the problem was how to get out of the other fellow's gunsight and get him in yours. Loops, Immelmann turns, and variations thereon abounded. Shown in Figures A.1 and A.2 are the loop and the hammerhead, or stall, turnaround.

Before you jump into your four-seat touring airplane, however, be advised that aerobatics are illegal unless performed in an aircraft rated for the maneuvers you want to try. And if you've never done such maneuvers you'd be well advised to purchase some dual beforehand. (Even the semi-aerobatic maneuvers required for the Commercial certificate hardly count.)

There are any number of reasons for taking up the sport, not least the confidence that comes from knowing you can handle your airplane from unusual attitudes. As a beginner you'll probably start work in a tail dragger—itself a useful exercise, since you'll learn how to handle a different style machine. Most people here start with

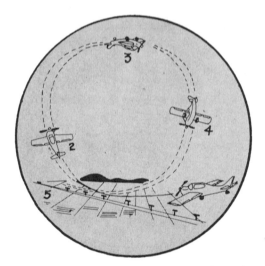

FIG. A.1.

For most people aerobatics means looping the loop. Perhaps the second most rudimentary of maneuvers—the dutch roll is the first—the loop is entered with wings level from a gentle dive to bring the aircraft up to 20 percent–plus normal cruise. If you throttled back to govern prop speed in the dive, you'll add this back as you ease back firmly on the stick. As you go inverted, ease back your Gs since you could high-speed stall over the top. You should keep positive G while inverted, and as you ease into the dive down the far side, throttle back and again avoid too high Gs as you pull up. The speed at which you come out should be the same as your entry speed.

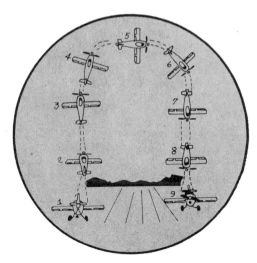

FIG. A.2.

The stall or hammerhead turn is a simple but highly effective basic maneuver. Using about 10 percent–plus normal cruise, pull the aircraft smoothly into a vertical climb, holding full power. As the ASI drops off the clock, kick full left or right rudder to make a 180-degree turn. After the briefest of dives, ease out into level flight. NOTE: airspeed buildup is dramatic—watch it! (Lt. Col. Art Medore's book **Primary Aerobatic Flight Training with Military Maneuvers**, from which these drawings are adapted, is required reading for anyone interested in aerobatic flight. Published by Ardot Enterprises, Inc., Dover, New Jersey, for $5.95.)

283

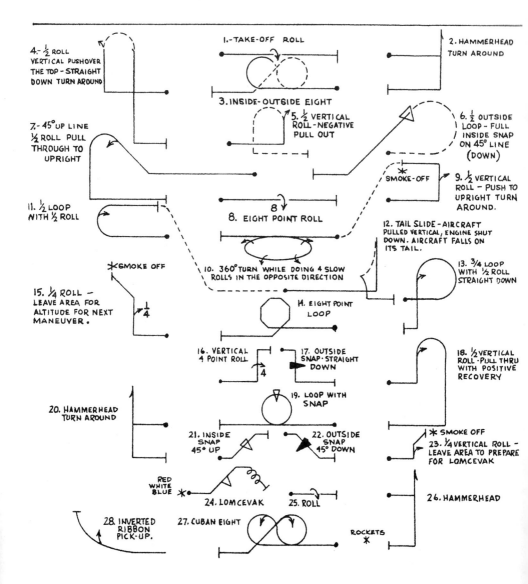

FIG. A.3.

Art Scholl's Aresti key.

284

FIG. A.4.

Like Eric-on-the-roof Carlson—the Saab team's top rally driver—Art Scholl spends much of his time upside down. Mostly it's in demonstrations in his Super Chipmunk Special at Air Shows, when not competing as a team member of the US International team. Seen here in his favorite Chipmunk, Art was shooting footage for **Jonathan Livingston Seagull**, the movie based on Richard Bach's best seller. Note the two Panavision cameras slung beneath the wings. The two brackets at the wing-tips are armed with smoke bombs in demonstration flights.

the Citabria, the modified version of the Champ (Aeronca), a pleasant if heavy machine to handle.

Your instructor will introduce you fairly slowly to each maneuver since the muscle building that goes on in a Citabria is considerable. Dutch rolls—where you prod one rudder and then the other, trying to harmonize it all together—develop the thighs rather well. Rolls develop the biceps, and eight point rolls—if you can hold each point—do wonders for your coordination and the muscles of the lower back.

Not all airplanes that perform aerobatically are like this, of course. But the advantage is that if you can handle even only halfway decently a Citabria (*airbatic* spelled backwards), you'll be much better able to command an aircraft with lighter inputs.

It does require some effort to become reasonably good, but the effort is certainly rewarded in the knowledge that you are that much better a pilot than those who have never attemped to turn the blue where the green is.

285

In competitions and in demonstration flights, a pilot has to know how to gather the various maneuvers he has to perform together. The essence of good aerobatics is the flow between one maneuver and the next, so that as one is ended the aircraft is in the right position to begin the next. To help pilots do this, the Spanish Count Aresti invented a shorthand, known as the Aresti key, to describe aerobatic maneuvers.

Shown in Figure A.3 is the Aresti key used by Art Scholl, one of the nation's most experienced aerobatic pilots. A member of the U.S. team, Scholl flies a DHC-1A "Chipmunk" trainer (Fig. A.4), made by De Haviland Aircraft of Canada, which he has modified for aerobatic performances. (In international competition he uses a Pitts Special).

Modifications include removal of some of each wing, the reduction of wing dihedral, and the fitting of new wing tips; increasing area and travel of rudder and fin; replacing fixed gear with retractable landing gear—and converting from two-seater to single place. (The extra room comes in handy for stowing his gear when he is flying between air shows). There is a fuel system that permits inverted flight, plus an oil-smoke system so that people can follow the line the airplane makes through the sky. Lastly, the original Gipsy Major engine has been replaced with a Lycoming GO-435-C2B, 260 hp motor.

As modified, the aircraft has a maximum dive speed of 250 mph and a stall speed of 65 mph. Maximum rate of climb is 2,500 fpm with a service ceiling of 23,500 feet.

When he's not traveling and demonstrating his prowess in flight around the country and abroad, Scholl provides aerobatic training at his own school in California.

Abbreviations

ACV	air-cushioned vehicle
ADF	automatic direction finding
AGL	above ground level
A/H	artificial horizon
AIM	Airman's Information Manual
Airmet	Advisory for Light Aircraft
ALPA	Airline Pilots Association
AOPA	Aircraft Owners and Pilots Association
A&P	Airframe & Powerplant (mechanic rating)
ASEL	airplane single-engine land
ASI	airspeed indicator
ATC	Air Traffic Control
ATIS	Automatic Terminal Information Service
ATR	air transport rating
DF	Direction Finder
DG	directional gyro
DME	Distance Measuring Equipment
EAA	Experimental Aircraft Association
EHF	Extremely High Frequency
ELT	emergency locator transmitter
EMDO	Engineering Manufacturing District Office (of FAA)
ETA	estimated time of arrival
EWAS	en route advisory service
FAA	Federal Aviation Administration
FAR	Federal Aviation Regulations
FBO	Fixed Base Operator
FCC	Federal Communications Commission

fpm	feet per minute
FSS	Flight Service Station
GADO	General Aviation District Office (of FAA)
gph	gallons per hour
IFR	Instrument Flight Rules
ILS	Instrument Landing System
LOM	Locator, Outer Marker
mph	miles per hour
MSL	Mean Sea Level
Notam	Notice to Airmen
NPA	National Pilots Association
NTSB	National Transportation Safety Board
OBS	Omnibearing Selector
PATWAS	Pilot's Automatic Telephone Weather Answering Service
PAR	Precision Approach Radar
Pirep	pilot report
RNAV	Area Navigation
rpm	revolutions per minute
RVR	runway visual range
SBA	Standard Beam Approach
SID	Standard Instrument Departure
Sigmet	Significant Meteorological Advisory
STAR	Standard Terminal Arrival Route
STOL	short takeoff and landing
TACAN	Tactical Air Navigation Aid
TBO	time before overhaul
TCA	Terminal Control Area
TWEB	Transcribed Weather Broadcast
UHF	Ultra High Frequency
VASI	Visual Approach Slope Indicator
VFR	Visual Flight Rules
VHF	Very High Frequency
VOR	Very High Frequency Omni-Directional Range
VORTAC	Combined VOR & TACAN System
VOT	VOR Test System
VTOL	vertical takeoff and landing; also V/STOL
WAC	World Areonautical Chart

Glossary

Aileron: The movable section of the outboard trailing edge of each wing that, operating asymetrically with the other, controls an airplane's movement about its roll axis.

Aileron yaw: The tendency of an airplane's nose to move in the direction of the dowward-moving aileron as a result of its drag being greater than the upward-moving ailerons.

Airfoil: A device—a wing, propeller, etc.—so shaped to take advantage of differential pressures created on its surfaces as a result of its movement through the air.

Airframe: The principal structure of an airplane, including its body or fuselage, wings, and tail, but not including its engine and other accessories.

Airspeed: The speed of an aircraft in relation to the air through which it is passing.

Airway: An air corridor designated by the FAA, controlled by Air Traffic Control, and provided with radio navigational beacons.

Airworthy: The state of an aircraft's being sound mechanically and safe to fly.

Altimeter: An aneroid barometer that is arranged to indicate the altitude of an airplane above sea level by measuring air pressure and giving the result in feet.

Altitude: Normally given in terms of height in feet above mean sea level; also given as height above the ground over which an aircraft is flying.

Angle of attack: The angle created between the airfoil of the wing and the wind relative to it.

Angle of incidence: The fixed angle relating to the position of an airplane's wing and the theoretical longitudinal axis of the aircraft.

Annual inspection: The inspection that must be carried out every twelve months by an appropriately certificated mechanic to verify the condition of airworthiness of an aircraft.

Approach: The final stage of flight in which an airplane is brought to a landing. The approach sector may extend for thirty or forty miles under ATC; the approach to the runway is then called final approach.

Attitude: Describes an airplane's position in relation to the horizon.

Axis, axes: Three theoretical lines extending fore and aft, vertical, and side to side through an airplane's center of gravity. These are more normally referred to as roll or longitudinal axis, yaw or vertical axis, and pitch or lateral axis, respectively.

Bank: The tilting or rolling of the airplane about the longitudinal axis to initiate a turn.

Base leg: The penultimate sector of the landing pattern, which lies at right angles to the final approach.

Carb heat: A control in the instrument panel that permits warmed air to be fed to the carburetor to remove or prevent the formation of ice.

Ceiling: The height of the base of the cloud cover above the ground. Occasionally quoted as altitude Mean Sea Level (MSL).

Center of gravity: Usually abbreviated CG or C of G, this is the theoretical balance point of an aircraft at which its three axes meet.

Checklist: An itemized account or written list of operating instructions. The safest way to ensure that you are operating an aircraft correctly is to follow the checklist.

Chord: The distance through a wing (or airfoil) measured as a straight line from leading to trailing edge.

Cockpit: The area within the fuselage of an airplane in which the pilot and crew sit and from which the aircraft is controlled.

Control stick, column: Early aircraft (and most European airplanes to this day) used a vertical stick or rod mounted on the floor, which was connected by cables and levers to the airplane's control surfaces. Most U.S. aircraft use a yoke or semiwheel attached to a column, mounted through the instrument panel and which is similarly connected to the control surfaces. A stick usually gives much better control response, while the yoke looks more like an airliner or automobile control wheel.

Control surface: Any moveable parts of an aircraft's wings or tail assembly used for maneuvering the aircraft.

Cross-control: A method of inducing a slip, either to lose height or to maintain track in a crosswind, in which the left aileron is used with the right rudder or vice versa. When cross-controlling, pilots pay close attention to their angle of attack, as a stall in this condition will normally result in a spin.

290

Cross-country: A flight between airports more then twenty-five nautical miles apart.

Crosswind: A condition in which the wind blows at an angle to an airplane's direction of flight.

Crosswind leg: That portion of the traffic pattern flown at right angles to the runway—directly opposite to the base leg.

Downwind: Traveling in the same direction as the wind.

Downwind leg: That portion of the traffic pattern in which an airplane is traveling downwind and parallel to the runway.

Drag: The resistance of air to the flight of an airplane opposite in direction to its motion.

Drift: The movement away from the chosen flight path of an airplane caused by a crosswind.

Dual, dual instruction: Flight training in an aircraft with a certified flight instructor.

Elevator: A moveable horizontal control surface that controls an airplane's angle of attack by raising or lowering the nose in flight.

Federal Aviation Administration: An administration within the Department of Transportation with responsibility to supervise, regulate, administer, and encourage aviation.

Final approach: See approach.

Flaps: A movable section of the inboard trailing edge of a wing that is hinged so that when it is lowered and moved outward it provides additional lift during slow flight. Fully extended, most flaps increase drag considerably, permitting steep descents and relatively low landing speeds.

Flare: The checking of an aircraft's downward descent, in which the nose is gently raised to level the airplane's flight path and the aircraft is held off the ground to produce a smooth landing.

Flight surfaces: Usually refers to such airfoil sections as wings and stabilizers; may also include vertical fins.

Forced landing: Landing an aircraft because of unforeseen circumstances. A precautionary landing, in which the pilot elects to land for some reason, is not a forced landing, which usually implies circumstances beyond the control of the pilot.

Fuselage: The main portion of an airframe, to which the wings and tail assembly are attached. The fuselage normally houses the engine, cockpit, passenger accommodations, and controls.

Glide: The condition of flight in which the forward motion of an airplane is maintained by trading altitude for forward speed by not using the engine and propeller.

Glide ratio: A figure that relates the forward distance an airplane will travel to the amount of altitude lost to maintain flight by gliding. Typically, a modern powered trainer has a glide ratio of approximately 11:1—that is, if it is one mile high it can travel eleven miles forward.

Ground loop: The inadvertent turning of an airplane on the ground, either while taxiing or landing, through 90 degrees at some speed, usually resulting in damage to either wing tip or landing gear. Nose-wheel aircraft, although not completely immune to ground looping, don't do it very often.

Ground speed: The speed of an airplane in relation to the ground over which it is flying. Ground speed relates to distance traveled over the ground; airspeed relates to speed relative to the air through which the airplane is flying.

Headwind: A wind blowing from some point in front of an aircraft, thereby slowing its ground speed.

Landing: Returning an aircraft from flight to a stop on the ground.

Landing gear: The wheels, braking mechanism, struts, and so on, which support an airplane when on the ground. The gear may be retracted during flight in certain aircraft, which are thus frequently referred to as retractables.

Lateral axis: See axis, axes.

Leading edge: The forward edge of any airfoil or flight surface.

Letdown: A descent, usually prior to landing, from cruise altitude.

Lift: The upward differential force created by the passage of an airfoil through the air produced by the wings of the aircraft, which keeps it up in flight.

Logbook: A pilot's record of flights made.

Longitudinal axis: See axis, axes.

Magneto: A specially designed generator used to supply the high-voltage current to an engine's spark plugs; also termed mag.

Mixture control: A control in the instrument panel with which the pilot adjusts the fuel/air mixture to the engine combustion chambers via the carburetor or fuel injector.

Nosewheel: A forward wheel ahead of the main landing gear that assists in supporting a tricycle-geared airplane's weight on the ground. It is frequently steerable, being interconnected with the rudder pedals.

292

Overcontrol: The tendency to move the airplane's controls too exuberantly or more than is needed.

Overshoot: To fly beyond a chosen point inadvertently or even to taxi beyond such a point.

Pattern: The rectangular flight path around an airfield, intended to provide an orderly flow of traffic arriving at and departing from the field. Sometimes referred to as traffic pattern.

Pitch: Refers to the nose-up or nose-down attitude of an airplane. See also axis, axes.

Pitot, Pitot tube: The tube projecting into the airstream that provides the source of ram-air pressure and that is correlated by the airspeed indicator with static air to produce airspeed indication. A heating device is usually supplied (as an option) to avoid its icing up.

Preflight inspection: The inspection conducted on the ground by the pilot-in-command of his airplane prior to any flight.

Propeller: A pair or more of airfoiled blades attached to an airplane engine's crankshaft, except in geared engines. May be wood, wood with metal facings on the leading edge, or metal. Propellers may have two, three, four, or even five blades.

Propeller wash: The turbulent air driven backward in a corkscrew-like motion by a moving propeller.

Pusher: An aircraft design in which the propeller is mounted behind the wing usually at the aft end of the fuselage. More efficient and quieter than the normal tractor-type arrangement.

Ram air: Used for cooling the engine, for pressurizing the ASI, and for cabin ventilation when the aircraft is in motion.

Roll out: To level the wings following a turn or bank.

Rudder: The movable vertical control surface at the rear of the vertical fin in an airplane's tail assembly, which controls the movement about the vertical axis of an airplane. Operated by the foot pedals in the cockpit.

Runup: The check made of the engine, magnetos, propeller (constant speed or variable pitch), and instruments prior to taking off on any flight.

Skid: A sideways motion of an airplane resulting from using too much rudder in relation to the ailerons.

Slip: A sideways motion of an airplane resulting from banking the airplane in one direction while applying rudder in the opposite direction. See cross-control.

Slipstream: See propeller wash.

Slow flight: Maintaining level flight at a speed above but close to the stall speed of an aircraft.

Solo: Flying an aircraft with only the pilot on board. Refers especially to a student's first flight on his own.

Span: Generally refers to wingspan, the distance from wingtip to wingtip.

Spin: The rotation of an airplane which is maintained in a stall by holding up elevator and caused to rotate by holding full left or right rudder. Used as a training maneuver. To stop a spin, apply full rudder in the opposite direction to the rotation and ease the control wheel forward gently until spinning stops, at which point rudder must be centralized at once.

Spiral: A highly banked, descending turn that when prolonged leads to redline airspeed; known also as a graveyard spiral.

Stability: The ability of an airplane to maintain itself in stable level flight without control correction.

Stabilizer: Particularly that horizontal non-movable flight surface forward of the elevator. On some aircraft the entire two sections move; this single surface is called a stabilator. The vertical fin is also sometimes known as the vertical stabilizer.

Stall: The condition in which, by reason of excessive angle of attack, the airflow over the wings is incapable of producing lift sufficient for the aircraft to maintain flight. In such conditions altitude is lost (some modern aircraft will not properly stall—they merely develop excessive sink) and stability is impaired. Usually the nose will drop as the aircraft attempts to fly at a more realistic angle of attack.

Static: May refer to atmospheric interference in radio communication, but more usually refers to static air pressure within the cabin or to undisturbed air at some point along the fuselage.

Stick: See control stick, column.

Tachometer: A gauge that provides information about engine rpm. Sometimes called a tach.

Tail wheel: The third supporting wheel found on older airplanes beneath the tail. Tail-wheel aircraft are occasionally referred to as tail-draggers by nosewheel drivers, who in turn are referred to as milk-stool pushers. Considerable argument (almost as much as between high-wing and low-wing enthusiasts) has developed between the two.

Tailwind: A wind blowing from a position astern an aircraft, thus increasing its groundspeed. A crosswind may have a tailwind component.

Takeoff: The process of bringing an airplane into flight from a stationary position on the ground.

Taxi: Maneuvering the airplane on the ground under its own power.

294

Taxiways: Paths or roads at airports specifically for the use of aircraft while maneuvering on the ground.

Thrust: Forward motion.

Torque: Usually refers to the tendency of an aircraft to turn in the direction opposite to its propeller rotation.

Touch-and-Go: Refers to the regime in which a student pilot (or any other pilot) is doing pattern work and in which immediately upon landing the carburetor heat is switched off, the throttle opened, and the aircraft taken off again. Circuits is the English term.

Tractor: Slang expression to describe front mounted engine and propeller.

Tricycle gear: See landing gear and nosewheel.

Trim tab: A very small hinged section of a control surface connected to a trim wheel within the cockpit and designed to reduce surface loading at the controls.

Undershoot: To fall short of a designated or desired point.

Weathervane: The tendency of an aircraft on the ground, especially on floats, to face into the wind as a result of wind pressure on the tail. Also called weathercocking.

Wind sock: A large fabric tubular sleeve attached to a swiveling metal hoop and mounted on a pole close to the runway. The light-weight material used will extend and turn indicating wind direction in light winds and will extend full out at 20 or 30 knots, depending on the fabric used.

Yaw: A situation in which the airplane's nose is not aligned longitudinally with its flight path.

Bibliography

Surprisingly, there is not a vast bibliography on the subject of flying, and much of what there is is out of print. The following is a selected reading list that covers the fundamentals of flight, airmanship, navigation, weather, and aviation rules and regulations. In addition, a short list of books of general and historical interest is also included.

FAA publications are obtainable from the U.S. Government Printing Office, Washington, D.C. 20402. They include a series of brief dockets known as VFR Exam-O-Grams, which form a question-and-answer series to the written part of the private pilot certificate.

Aero Products Research Inc. markets what they describe as "a complete management and administrative system" of pilot training, known as "Airlearn." This system is similar to the integrated training courses offered by Beech, Cessna, Grumman American, and Piper Aircraft companies, to those of Sanderson Films Inc., and to those produced by Jeppesen & Company. Each system provides complete flight-training manuals.

FUNDAMENTALS OF FLIGHT

A Flight Guide for Pilots and Flight Instructor, by Al Guthrie. Al Guthrie, P.O. Box 241, San Carlos, California 94070.

Airguide Manual: Flight Training Handbook. Airguide Publications, Long Beach, California.

Airplane Mechanics Manual, A Zweng Manual. Pan American Navigation Service Inc., North Hollywood, California, 91601.

Campbell's Pilot Rating Guides. Aviation Book Company, Glendale, California.

Computer Guide, by Frank K. Smith. Sports Car Press, New York, N.Y.

Facts of Flight. FAA Publication 1.8:F64/3/963.

Flight Facts for Private Pilots, by Merrill E. Tower. (1960) Aero, Fallbrook, California.

Flight Maneuvers: Complete Programmed Course (2nd Ed., 1969), Aero Products Research, Inc., Aviation Education Dept., Los Angeles, California, 90045.

Flight Planning Guide for Pilots by Larry Reithmaier. Aero, Fallbrook, California.

Flight Training Handbook. FAA Publication AC 61-16.

Flight Training Syllabus. Institute of Aeronautical Safety, Los Angeles, California, 90045.

296

Fundamentals of Flight. Aviation Book Co., Glendale, California.
Ground School Workbook (3rd Ed., 1972), by Betty Hicks. Iowa State University Press, Ames, Iowa.
Modern Airmanship (4th Ed., 1971), ed. by Neil D. Van Sickle. Van Nostrand Reinhold Company, New York, N.Y.
New Private Pilot, A Zweng Manual. Pan American Navigation Service, North Hollywood, California, 91601.
Path of Flight. FAA Publication 1.8: F64/2/963.
Pilot Training and Reference Manual Kit (13 pts., 1973), Aviation Book Company, Glendale, California.
Practical Manual for E6B Computer, A Zweng Manual. Pan American Navigation Service, North Hollywood, California.
Private Pilot Course. Jeppesen & Co., Denver, Colorado.
Private Pilot's Handbook of Aeronautical Knowledge. FAA Publication AC 61-23.
Private Pilot: Complete Programmed Course (4th Ed., 1973). Aero Products Research, Aviation Education Dept., Los Angeles, California, 90045.
Private Pilot's Guide (1972), by Larry Reithmaier. Aero, Fallbrook, California.
Private Pilot Study Guide (1970), by Leroy Simonson. Aviation Book Company, Glendale, California.
Realm of Flight. FAA Publication 1.9: F64/963.
Stick and Rudder (1944), by Wolfgang Langewiesche: McGraw-Hill Inc., New York, N.Y.
Student Pilot's Flight Manual (4th Ed., 1973), by William K. Kershner. Iowa State University Press, Ames, Iowa.
Student Pilot Guide. FAA Publication AC 61-12C.
Using Your Navigation Computer. Aviation Book Company, Glendale, California.
Your FAA Flight Exam (1968), by Bob Smith. Sports Car Press, New York, N.Y.
Your Pilot's License (rev. Ed., 1960), by Joe Christy & Clay Johnson, Sports Car Press, New York, N.Y.

REGULATIONS

Airman's Information Manual. FAA Publication.
Airman's Information Manual (1973). Aero Fallbrook, California.
Airman's Information Manual. Special Pilot Training Edition (1973), edited by Walter P. Winner. Aviation Book Company, Glendale, California.
Federal Aviation Regulations for Pilots (rev. Ed., 1973). Aero, Fallbrook, California.
Federal Aviation Regulations For Pilots (8th Ed., 1973), Pts. 1, 61, 91, 430; ed. by Leroy Simonson. Aviation Book Company, Glendale, California.
Federal Aviation Regulations (Pts. 1, 61, 91, 93, 430): *Handbook for Pilots*, ed. by Ellen Casselbury. Aircraft Owners and Pilots Association, Washington, D.C.
Federal Aviation Regulations for Pilots, A Zweng Manual. Pan American Navigation Service, North Hollywood, California.
Pilots and Aircraft Owners Legal Guide, by J. C. White. Aircraft Owners and Pilots Association, Washington, D.C.

AIRMANSHIP

Advanced Pilot's Flight Manual (3rd Ed., 1970), by William K. Kershner. Iowa State University Press, Ames, Iowa.

Advanced Private and Commercial Course: Jeppesen & Co., Denver, Colorado.

Aircraft Weight and Balance Control (rev. Ed., 1967), by Henri G. D'Estout. Aero, Fallbrook, California.

Airmanship After Solo, (1968) by Don Downie. Sports Car Press, New York, N.Y.

Air Transport Pilot, A Zweng Manual. Pan American Navigation Service, Inc., North Hollywood, California.

The Art and Technique of Soaring (1971), by Richard A. Wolters. McGraw-Hill Inc., New York, N.Y.

The Cessna 150 Aerobat Training Manual, Cessna Aircraft Co., Wichita, Kansas 67201.

Commercial Pilot Study Guide (1973), by Leroy Simonson. Aviation Book Company, Glendale, California.

Flying with Floats: A Guide to Seaplane Operations and Techniques (1966), by Alan Hoffsommer. Pan American Navigation Service, North Hollywood, California.

Helicopter Pilot Ratings, A Zweng Manual. Pan American Navigation Service, North Hollywood, California.

Instrument Flying (1972), by Richard L. Taylor. The Macmillan Company, New York, N.Y.

Instrument Flying Guide, by Robert T. Smith. Sports Car Press, New York, N.Y.

Instrument Flying Handbook, FAA Publication AC 61-27A.

Instrument Pilot (Airplane), Flight Test Guide. FAA Publication AC 61-17A.

The Instrument Rating (21st Ed., 1973), A Zweng Manual. Pan American Navigation Service, North Hollywood, California.

Joy of Soaring: A Training Manual (1969), by Carle Conway. Aviation Book Co., Glendale, California.

Modern Aerobatics and Precision Flying (rev. ed., 1970), by Harold Krier. Sports Car Press, New York, N.Y.

Multiengine Flying, by Alice Fuchs. Crown Publishers, Inc., New York, N.Y.

Multiengine Airplane Rating (4th Ed., 1973), A Zweng Manual. Pan American Navigation Service, North Hollywood, California.

New Commercial Pilot (19th Ed., 1972), A Zweng Manual. Pan American Navigation Service, North Hollywood, California.

On Silent Wings (1970), by Don Dwiggins. Grosset & Dunlap, New York, N.Y.

Pilot's Handbook of Instrument Flying (1969), by Larry Reithmaier. Aero, Fallbrook, California.

Primary Aerobatic Flight Training: How to Fly Beginning and Advanced Aerobatics, by Art Medore. Aviation Book Co., Glendale, California.

Roll Around a Point, by Duane Cole. Ken Cook Company, Milwaukee, Wisconsin.

Visualized Flight Maneuvers Check Ride Check List (1971). Aviation Book Co., Glendale, California.

Weather Flying (1970), by Robert N. Buck. The Macmillan Company, New York, N.Y.

NAVIGATION

Cockpit Navigation Guide, by Don Downie. Sports Car Press, New York, N.Y.

Dead Reckoning Navigation. Aviation Book Company, Glendale, California.

Pilot's Handbook of Navigation (1967), by James C. Elliott & Gene Guerny. Aero, Fallbrook, California.

Using Aeronautical Charts. Aviation Book Company, Glendale, California.

WEATHER

Air Masses and the Weather. Aviation Book Company, Glendale, California.

Aviation Weather. FAA Publication AC 00-6.

Flying and the Weather. Aviation Book Company, Glendale, California.

Interpreting Teletype Weather Data. Aviation Book Company, Glendale, California.

Introduction to Meteorology (3rd Ed., 1969), by Sverre Petterssen. McGraw-Hill Inc., New York, N.Y.

Meteorology for Glider Pilots, by C.E. Wallington. John Murray, Ltd., London, England.

Pilot's Handbook of Weather (1966), by Gene Guerny & Joseph A. Skiera. Aero, Fallbrook, California.

Watching for the Wind: The Seen and Unseen Influences on Local Weather (1967), by James G. Edinger. Doubleday, New York, N.Y.

Weather Briefing Guide for Pilots, by Larry Reithmaier. Aero, Fallbrook, California.

Weather Flying (1970), by Robert N. Buck. The Macmillan Company, New York, N.Y.

HISTORICAL

The Air Devils, by Don Dwiggins. J.B. Lippincott, Philadelphia, Pa.

Air Facts and Feats (1971), comp. by Francis K. Mason & Martin C. Windrow. Doubleday, New York, N.Y.

Before the Eagle Landed: The Saga of Aviation Told by Those Who Were There (1970), by the editors of the Air Force Times: Robert B. Luce Inc., Washington, D.C.

Before the Wrights Flew, by Stella Randolph. G.P. Putnam's Sons, New York, N.Y.

Bold Men, Far Horizons, by Herbert Molloy Mason, Jr. J. B. Lippincott, Philadelphia, Pa.

Curtiss Standard JN4-D Military Tractor Handbook. Aviation Publications, Box 123, Milwaukee, Wisc.

Command the Horizon, by Page Shamburger and Joe Christy. A. S. Barnes & Co., Cranbury, N.J.

The Dream of Flight: Aeronautics from Classical Times to the Renaissance (1972), by Clive Hart. Winchester Press, New York, N.Y.

Duel of Eagles (1973), by Peter Townsend. Simon & Schuster, New York, N.Y.

The First to Fly, by Sherwood Harris. Simon & Schuster, New York, N.Y.

Flying Boats and Seaplanes Since 1910, by Kenneth Munson. The Macmillan Company, New York, N.Y.

The Great Planes (1970), by James Gilbert: Grosset & Dunlap, New York, N.Y.

The American Heritage History of Flight, by the editors, American Heritage Library Publishing Co., Inc., New York, N.Y.

How We Invented the Airplane, by Orville Wright. David McKay Company, New York, N.Y.

Milestones of the Air: Jane's 100 Significant Aircraft (1971), by John W. R. Taylor and H. F. King. McGraw-Hill Book Company, New York, N.Y.

Mosquito, by Joe Holliday. Doubleday, New York, N.Y.

Mr. Piper and His Cubs, by Devon Francis. Iowa State University Press, Ames, Iowa.

Of a Fire on the Moon (1971), by Norman Mailer. New American Library.

BIBLIOGRAPHY

Ryan the Aviator, by William Wagner and Lee Dye. McGraw-Hill Inc. New York, N.Y.

The Pilot Maker (1970), by Lloyd L. Kelly, as told to Robert B. Parke. Grosset & Dunlap, New York, N.Y.

The Saga of the Air Mail (1968), by Carroll V. Glines. Van Nostrand Reinhold Company, New York, N.Y.

Sir George Cayley's Aeronautics 1786–1885, by C. H. Gibbs-Smith. Soaring International.

Smithsonian Annals of Flight (7 vols.). U.S. Government Printing Office, Washington, D.C.

Sopwith—the Man and His Aircraft, by Bruce Robertson. Air Review Ltd., Letchworth, Hertfordshire, England.

The Spirit of St. Louis (1953), by Charles A. Lindbergh. Charles Scribner's Sons, New York, N.Y.

Veteran and Vintage Aircraft (1971), by Leslie Hunt. Taplinger Publishing Co., New York, N.Y.

The Wright Brothers (1966), by Fred C. Kelly: Ballantine, New York, N.Y.

GENERAL INTEREST

Aircraft Dope and Fabric Guide (1970), by Ruth and Warren Spencer. Sports Car Press, New York, N.Y.

All About Flying (1964), by Barry Schiff. Aviation Book Co., Glendale, California.

America's Flying Book (1972), by the editors of *Flying* magazine. Charles Scribner's Sons, New York, N.Y.

Antoine de Saint-Exupery (1970), by Curtis Cate. G. P. Putnam's Sons, New York, N.Y.

Biplane (1966), by Richard Bach. Harper & Row, New York, N.Y.

The Complete Book of Sky Sports (1970), by Linn Emrich. The Macmillan Company, New York, N.Y.

Fate is the Hunter (1972), by Ernest K. Gann. Simon & Schuster, New York, N.Y.

The Human Factor in Aircraft Accidents (1969), by David Beaty. Stein & Day, New York, N.Y.

I'd Rather Be Flying: Instrument and Twin Engine Flying for the Weekend Pilot (1962), by Frank Kingston Smith. Random House, New York, N.Y.

I Live to Fly (1970), by Jacqueline Auriol. E. P. Dutton & Co., Inc., New York, N.Y.

Jane's All the World's Aircraft (1973, 1974), edited by John W. R. Taylor. McGraw-Hill Book Company, New York, N.Y.

Night Flight, by Antoine de St. Exupery. New American Library.

North to the Orient, by Anne Morrow Lindbergh. Harcourt, Brace & Jovanovich, New York, N.Y.

Pilot's Guide to an Airline Career (3rd Ed., 1971), by W. L. Traylor. Aviation Book Company, Glendale, California.

Rickenbacker (1967), by Edward V. Rickenbacker. Prentice-Hall, Englewood Cliffs, N.J.

Round the Bend, by Nevil Shute. Ballantine, New York, N.Y.

The Single-Engine Beechcrafts (1970), by Joe Christy. Sports Car Press, New York, N.Y.

Southern Mail (1972), by Antoine de St. Exupery. Quinn & Boden, New York, N.Y.

Stranger to the Ground (1972), by Richard Bach. Harper & Row, New York, N.Y.

This is EAA, by Duane Cole. Duane Cole, 201 Lester Street, Burleson, Texas 76028.

Those Incomparable Bonanzas, by Larry A. Ball. Larry A. Ball, 8407 Peach Tree Lane, Wichita, Kansas, 67207.

The Tiger Moth Story, by Alan Bramson and Neville Birch. Air Review Ltd., Letchworth, Hertfordshire, England.

Ulysses Airborne (1971), by Mauricio Obregon. Harper & Row, New York, N.Y.

Used Plane Buying Guide (rev. ed., 1968), by James R. Trigg. Sports Car Press, New York, N.Y.

Wind, Sand and Stars, by Antoine de St. Exupery. Harcourt, Brace & Jovanovich, New York, N.Y.

Winged Legend: The Story of Amelia Earhart (1970), by John Burke. G. P. Putnam's Sons, New York, N.Y.

Your Future as a Pilot, Captain Kimball J. Scribner: Arco Publishing Company, Inc., New York, N.Y.

MAGAZINES

Air Facts (monthly). 110 E. 42nd St., Suite 1419, New York, N.Y. 10017.

Air Progress (monthly). Petersen Publishing Company, 8490 Sunset Boulevard, West Hollywood, Los Angeles, Calif. 90069.

AOPA Pilot (monthly). P.O. Box 580, Washington, D.C. 20014.

The Aviation Consumer (bi-monthly). Belvoir Publications, Inc., 6 W. 37th St., New York, N.Y. 10018.

Aviation Daily. Ziff-Davis Publishing Company, 1156 15th St. N.W., Washington, D.C. 20005.

Business & Commercial Aviation, Ziff-Davis (see *Flying*).

Flying (monthly). Ziff-Davis Publishing Company, 1 Park Avenue, New York, N.Y. 10016.

1974 Flying Annual & Pilot's Guide. Ziff-Davis (See *Flying*).

1974 Invitation to Flying. Ziff-Davis (See *Flying*).

1974 Instrument Flying Guide. Ziff-Davis (See *Flying*).

Plane & Pilot (monthly). Werner & Werner, 631 Wilshire Boulevard, Santa Monica, Calif. 90406.

Private Pilot (monthly). Macro-Com Corp. 291 South LaCienega Boulevard, Beverly Hills, Calif. 90211.

Soaring. Soaring Society of America, P.O. Box 66071, Los Angeles, Calif. 90066.

Sport Aviation. Experimental Aircraft Association, P.O. Box 229, Hales Corner, Wisc., 53130.

Index

Phugoid oscillation, 44
Pilot reports. *See* Pireps
Pilot training centers, 4
Pilot's Automatic Telephone Weather Answering Service. *See* PATWAS
Piper, 140, 142–45
  performance, 145
  specifications, 145
Piper Cubs, 143, 152–54, 259
Pireps, 88
Piston Engines, 101–6
Pitot tube, 12
Power-off emergency, 66–70
Precession, 17
Preflight checklists, 15
Private certificates, 114
  instrument ratings and, 116
Procedure turns, instrument flying and, 253–55
Progressive maintenance, 144
Projection, sectional charts and, 91, 92
Propellers, 8–9
  cruise or climb, 35
  functioning of, 34–35
  maintenance, 212
  torque in, 43
Psychological testing of pilots, 122–23

Radios
  in airplanes, 217–28
  congestion and, 217–18
  costs, 221–28
  frequency allocations, 233
  ionosphere and, 79
  licensing, 232–33
  navigation systems, 117
  phonetic alphabet, 236
  secondhand equipment, 222
  sending numbers, 236
  using, 235–37
  waves and wavelengths, 218–21
Rate of descent, landing and, 59
Ratings, 113–35
  multi-engine, 118–20
  seaplane, 120
  See also Instrument rating.
Renting, costs of, 201, 204–5
Roll control, 39–40
Rotary-wing homebuilts, 176–77
Rudders, 10
  turning and, 40–41
Runway visual range reports, 88

Sailplanes, homebulit, 177–78
Salaries, corporate aviation, 127–30
Salesmen. *See* Aircraft salesmen
Schools. *See* Training schools
Seaplane ratings, 120

Sectional charts, 91–92
  magnetic variations on, 90
Senecas, pilot salaries, 128
Short takeoff and landing aircraft. *See* STOL
Sigmets, 86, 87, 88
Sign-towing, 130
Significant Meteorological Advisories. *See* Sigmets
Simulators, use of, 4
Skywriting, 130
Special Airworthiness Certificate, 159
Spins, 30, 53–56
Sports flying, 134
Squall lines, 82
Stabilizers, 11
Stall/spin syndrome, 65
Stall-warning devices, 11
Stalls, 30
  Cessna, 141, 150
  delaying the, 51
Stanine test, 123, 124
Station wagon, 192
Stationary fronts, 83
Steep turns, 76–77
Stinson voyager, 192
STOL aircraft
  air-cushioned landing gear, 266
  channel-wing design, 262
  general aviation and, 265
  lift needed, 264
  noise and, 265, 267
  performance, 259
  public acceptance of, 267
  space needed, 263
  technology of, 262
  uses, 268
Straight and level flight, learning, 30, 44
Stratiform clouds, 82
Stratosphere, 79
Student licenses, 113
Sturgeon air, 171–72
Survival equipment, 231–32

Tachometer, 19
Takeoffs, 61–64
  engine failure and, 64–66
Taxiing, 30
Taylorcrafts, 183–85
Telephone, 231
Test-piloting, 130
Throttles, 11, 19
Thunderstorms, 84
Time, 237
Tires
  costs, 202
  maintenance, 211
  prices, 209